JAMESTOWN'S

Number Power

Connie Eichhorn

JAMESTOWN PUBLISHERS

a division of NTC/Contemporary Publishing Group
Lincolnwood, Illinois USA

ISBN: 0-8092-2280-9

Published by Jamestown Publishers,
a division of NTC/Contemporary Publishing Group, Inc.,
4255 West Touhy Avenue,
Lincolnwood (Chicago), Illinois 60712-1975 U.S.A.

4 5 6 7 8 9 10 11 12 021 10 09 08 07

Contents

To the Student

This Number Power book is designed to help you build and use the basic math skills you need to handle measurement situations found in many work settings and in your everyday experiences. This practice also will help you improve your basic math ability.

Each section of *Measurement* provides settings where measurement skills are used. There are step-by-step examples and plenty of practice exercises to develop and reinforce your understanding and skill with measurements. Each section begins with a common work-centered illustration of measurement and measurement skills, and concludes with a review of all the skills covered up through that section. Throughout the book you have opportunities to apply your math skills in common, real measurement situations. Both customary units and metric units are presented throughout the book. Also, whole numbers, fractions, and decimals are used throughout the book. At the back of this book are an index, a glossary, and formulas, which are useful tools for quick review and helpful explanation.

The measurement skills you develop and practice through this book will be helpful in your work, at home, at school, and in other everyday situations.

Pre-Test
Math and Measurement Skills Inventory

1. $\begin{array}{r} 2{,}438 \\ +\ 6{,}351 \\ \hline \end{array}$

2. $\begin{array}{r} 53{,}476 \\ +\ 46{,}523 \\ \hline \end{array}$

3. $\begin{array}{r} 306{,}935 \\ +\ 292{,}042 \\ \hline \end{array}$

4. $28 + 46 + 37 =$

5. $216 + 492 + 347 =$

6. $5{,}643 + 975 + 1{,}605 =$

7. $4{,}738 + 921 + 16 + 1{,}204 =$

8. $\begin{array}{r} 426{,}528 \\ +\ 375{,}633 \\ \hline \end{array}$

9. $\begin{array}{r} 387 \\ -\ 142 \\ \hline \end{array}$

10. $\begin{array}{r} 4{,}052 \\ -\ 1{,}021 \\ \hline \end{array}$

11. $\begin{array}{r} 26{,}427 \\ -\ 13{,}215 \\ \hline \end{array}$

12. $8{,}746 - 2{,}413 =$

13. $204 - 186 =$

14. $27{,}005 - 18{,}927 =$

15. $3{,}049 - 956 =$

16. 427,361
   − 284,196

17. 53
   × 3

18. 46
   × 8

19. 91
   × 5

20. 6,102
   × 4

21. 27 × 35 =

22. 49
   × 30

23. 607
   × 18

24. 789 × 65 =

25. 8,306
   × 471

26. 18,475
   × 609

27. 619 × 200 =

28. 1,000
   × 347

29. 5)175

30. 7)2,940

31. 8)4,630

32. 28)4,936

33. 75)24,075

34. 4,780 ÷ 40 =

35. 29,000 ÷ 100 =

36. 4,670 ÷ 1,000 =

37. Reduce $\frac{12}{20}$.

38. Reduce $\dfrac{18}{24}$.

39. $\dfrac{1}{4} + \dfrac{1}{4} =$

40. $\dfrac{3}{8} + \dfrac{1}{4} =$

41. $\begin{array}{r} 2\frac{1}{2} \\ + \ 3\frac{1}{3} \\ \hline \end{array}$

42. $\begin{array}{r} 1\frac{3}{4} \\ + \ 4\frac{1}{2} \\ \hline \end{array}$

43. $\begin{array}{r} 6\frac{2}{3} \\ + \ 1\frac{1}{3} \\ \hline \end{array}$

44. $\dfrac{4}{5} - \dfrac{1}{5} =$

45. $\dfrac{2}{3} - \dfrac{1}{2} =$

46. $\begin{array}{r} 3\frac{1}{2} \\ - \ 2\frac{1}{4} \\ \hline \end{array}$

47. $\begin{array}{r} 6 \\ - \ 3\frac{2}{3} \\ \hline \end{array}$

48. $\begin{array}{r} 4\frac{1}{2} \\ - \ 1\frac{3}{5} \\ \hline \end{array}$

49. $\dfrac{1}{2} \times \dfrac{7}{10} =$

50. $2\frac{1}{3} \times 5 =$

51. $4\frac{1}{2} \times 3\frac{1}{3} =$

52. $\dfrac{1}{2} \times 2\frac{1}{2} =$

53. $2\frac{1}{4} \times \dfrac{2}{3} =$

54. $\dfrac{4}{5} \div \dfrac{1}{2} =$

55. $2\frac{1}{2} \div \dfrac{1}{2} =$

56. $1\frac{1}{4} \div 2 =$

57. $6\frac{1}{2} \div 1\frac{1}{4} =$

58. $5 \div 1\frac{1}{4} =$

59. 2.6
    + 4.8
----

60. 16.3
    + 3.6
----

61. 2.73 + 4.2 + 6.0 =

62. 3.05 + 0.28 + 4.37 =

63. 42.68
    + 39.842
----

64. 6.95
    − 3.74
----

65. 9.8
    − 4.64
----

66. 236.84 − 198 =

67. 206.6 − 48.75 =

68. 578.42
    − 229.56
----

69. 89.4
    × 17
----

70. 206
    × 4.05
----

71. 37.6
    × 0.01
----

72. 42.56
    × 100
----

73. 0.385 × 0.1 =

74. 496.35 × 2.6 =

75. 2.2)‾46.2‾

76. 0.01)‾3.5‾

77. 58)‾997.6‾

78. 0.46)‾.5152‾

79. 84.75 ÷ 100 =

80. 329 ÷ 1,000 =

81. If a plane has an average speed of 350 miles per hour, how long would it take to go 1,400 miles?

82. The Smith family spent $52 for electricity in April, $46 in May, $58 in June, and $66 in July. What was their average monthly bill for electricity for the four months?

83. Jai is paid $420 every two weeks. How much is he paid in one year?

84. Kylie bought a new sofa for $329. After she made a down payment of $79, how much did she owe for the sofa?

85. Jim's original weight was 206 pounds. If he lost 28 pounds and then gained 12 pounds, what is his current weight?

86. Jeni is buying material to make 8 costumes for a community production. If each costume uses $2\frac{3}{4}$ yards of fabric, how much fabric should she buy?

87. Josie started her paint job with four and a half gallons of paint. At the end of the day, she had $1\frac{3}{4}$ gallons left. How much paint did she use during the day?

88. Teresa bought a 30-yard spool of lace trim. She is decorating cloths, and each cloth takes $1\frac{1}{2}$ yards of lace. How many cloths can she decorate with the 30-yard spool?

89. Juan had $247.58 in his checking account. He wrote a check for $36.95 and then deposited $62.46. What is his new balance?

90. Tami bought ground beef for $1.59 per pound. If she paid $13.52, how many pounds of ground beef (to the nearest tenth) did she buy?

Answers are on pages 6–7.

Skills Inventory Answers

1. 8,789	**26.** 11,251,275	**51.** 15
2. 99,999	**27.** 123,800	**52.** $1\frac{1}{4}$
3. 598,977	**28.** 347,000	**53.** $\frac{3}{2} = 1\frac{1}{2}$
4. 111	**29.** 35	**54.** $1\frac{3}{5}$
5. 1,055	**30.** 420	**55.** 5
6. 8,223	**31.** 578 r 6 or 578.75	**56.** $\frac{5}{8}$
7. 6,879	**32.** 176 r 8	**57.** $5\frac{1}{5}$
8. 802,161	**33.** 321	**58.** 4
9. 245	**34.** 119 r 20 or 119.5	**59.** 7.4
10. 3,031	**35.** 290	**60.** 19.9
11. 13,212	**36.** 4.67	**61.** 12.93
12. 6,333	**37.** $\frac{3}{5}$	**62.** 7.7
13. 18	**38.** $\frac{3}{4}$	**63.** 82.522
14. 8,078	**39.** $\frac{1}{2}$	**64.** 3.21
15. 2,093	**40.** $\frac{5}{8}$	**65.** 5.16
16. 143,165	**41.** $5\frac{5}{6}$	**66.** 38.84
17. 159	**42.** $6\frac{1}{4}$	**67.** 157.85
18. 368	**43.** 8	**68.** 348.86
19. 455	**44.** $\frac{3}{5}$	**69.** 1,519.8
20. 24,408	**45.** $\frac{1}{6}$	**70.** 834.3
21. 945	**46.** $1\frac{1}{4}$	**71.** 0.376
22. 1,470	**47.** $2\frac{1}{3}$	**72.** 4,256
23. 10,926	**48.** $2\frac{9}{10}$	**73.** 0.0385
24. 51,285	**49.** $\frac{7}{20}$	**74.** 1,290.51
25. 3,912,126	**50.** $11\frac{2}{3}$	**75.** 21

76. 350	**81.** 4 hours	**86.** 22 yards
77. 17.2	**82.** $55.50	**87.** $2\frac{3}{4}$ gallons
78. 1.12	**83.** $10,920	**88.** 20 cloths
79. 0.8475	**84.** $250	**89.** $273.09
80. 0.329	**85.** 190 lb	**90.** 8.5 lb

Skills Inventory Chart

If you got fewer right than you should have, or if you cannot tell why your answers are not correct, you can review sections in the other Number Power books. There you will find explanations and more practice problems for each skill in this inventory.

Rework any problems you missed in the Skills Inventory. Then, when you are satisfied with your results, begin work in this Measurement book.

Problem Numbers	**Other *Number Power* Books**
1–36, 81–85	*Addition, Subtraction, Multiplication, and Division*
37–58, 86–88	*Fractions, Decimals, and Percents*
59–80, 89–90	*Algebra*

Section 1
Measurements

When Do We Use Them?

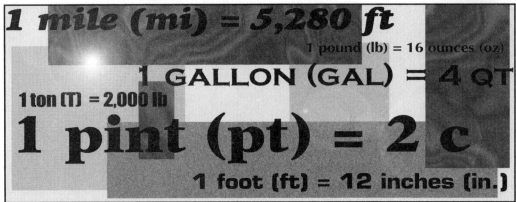

1 mile (mi) = 5,280 ft

1 pound (lb) = 16 ounces (oz)

1 GALLON (GAL) = 4 QT

1 ton (T) = 2,000 lb

1 pint (pt) = 2 c

1 foot (ft) = 12 inches (in.)

We use measurements when we describe people, places, and things. For example, a driver's license describes a person's height, weight, and age; and a building plan describes lengths, heights, angles, areas, and volumes.

We use measurements when we want to compare two things. For example, you can tell if one liquid is hotter than another by measuring the temperature of each one. Or, you can tell if one motorboat is faster than another by measuring the top velocity of each one.

We use measurements when we want to find differences between things. For example, to find how much gasoline you used on a trip, you could find the difference between the number of gallons you started with and the number of gallons you ended with. Or, to find the length of a car trip, you could subtract the start-of-trip odometer reading from the end-of-trip odometer reading.

All measurements have two parts, a number and a unit. The number may be a whole number, a fraction, or a decimal. So, when you use measurements you will be practicing your skills at adding, subtracting, multiplying, and dividing with different types of numbers.

In this book, you will learn about many different kinds of units, including units for weight, length, temperature, and other physical characteristics of objects. Some units, like ounces, pounds, feet, quarts, and gallons, are **Customary Units of Measurement** (sometimes called **Imperial units**). Other units, like grams, meters, and liters, are **Metric Units of Measurement** (sometimes called **International units**).

Think about your daily activities at work, at home, and at leisure. When do you use measurements?

Here are some of the questions you will be asked about measurements in this book.

- **What is an appropriate unit to use to estimate a measurement?**

- **How do you change a measurement from one unit to another unit?**

- **How do you add or subtract two measurements?**

- **How do you multiply or divide a measurement by a number?**

- **How do you apply measurements in everyday situations?**

As you use measurements, you will develop and review your math skills involving whole numbers, decimals, and fractions. Measurements are important for almost all jobs, for many daily activities, and for understanding and solving problems. In these lessons, you will investigate tools and scales for finding measurements. Also, you will apply those tools and scales using different kinds of units.

Career Corner

Many jobs require measurements. If a worker's tasks include building, covering, fitting, cooking, feeding, counting, installing, mixing, growing, or many other activities, then the job involves measurement. Also, people with jobs as testers, inspectors, and product graders rely on measurements. Here are some of the jobs that require measurement skills.

Length and Angles
real estate appraiser
bricklayer
electrician
tool-and-die worker

Weight and Temperature
HVAC technician (heating,
ventilation, and air conditioning)
meat cutter
welder

Capacity and Volume
boilermaker
counter attendant
fast-food worker
stationary engineer

Time and Velocity
air traffic controller
bus driver
truck driver
railroad engineer

Can you think of other jobs for any of these four categories?

Customary and Metric Units

Does your city or town have a watertower? If so, it might look like the tower at the right. To describe this water tower, we might say that it is 115 feet tall, weighs 16,000 pounds (when empty), and can hold 25,000 gallons of water.

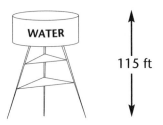

In this description, 115 feet is a measurement of *length*, 16,000 pounds is a measurement of *weight*, and 25,000 gallons is a measurement of *capacity*. Here is how some of the customary units for length, weight, and capacity are related to each other.

Length	Weight	Capacity
1 foot (ft) = 12 inches (in.)	1 pound (lb) = 16 ounces (oz)	1 cup (c) = 8 fluid ounces (fl oz)
1 yard (yd) = 36 in.	1 ton (T) = 2,000 lb	1 pint (pt) = 2 c
1 yd = 3 ft		1 quart (qt) = 2 pt
1 mile (mi) = 5,280 ft		1 gallon (gal) = 4 qt
1 mi = 1,760 yd		

Study the chart above. What is the abbreviation for each unit?

1. yards ____

2. tons ____

3. inches ____

4. pints ____

5. miles ____

6. fluid ounces ____

7. pounds ____

8. gallons ____

9. quarts ____

Fill in each blank with the correct equivalent measurement.

10. 1 c = ____ fl oz

11. 1 ft = ____ in.

12. 1 ton = ____ lb

13. 1 lb = ____ oz

14. 1 gal = ____ qt

15. 4 ft = ____ in.

16. 1 pt = ____ c

17. 1 mi = ____ yd

18. 3 gal = ____ qt

Name three things that are sold, measured, or packaged in the given unit.

19. miles _____

20. gallons _____

21. ounces _____

22. pounds _____

In metric units, the basic unit of length is the **meter**, the basic unit of weight is the **gram**, and the basic unit of capacity is the **liter**.

Most track-and-field races are measured in meters. A meter is a little longer than a yard. Some medicines are sold in grams. A gram is approximately the weight of a vitamin tablet. Many large soft-drink containers are measured in liters. A liter is a little more than a quart.

The metric system uses two important prefixes at the beginning of a unit. They are "milli-," which means "one thousandth of," and "kilo-," which means "one thousand times." Here is how the metric units are related to each other.

Length	Weight	Capacity
1 meter (m) = 1,000 millimeters (mm)	1 gram (g) = 1,000 milligrams (mg)	1 liter (L) = 1,000 milliliters (mL)
1 kilometer (km) = 1,000 m	1 kilogram (kg) = 1,000 g	1 kiloliter (kL) = 1,000 L

Study the chart above. What is the abbreviation for each unit?

23. milliliter _____   26. kilometer _____   29. gram _____

24. kilogram _____   27. liter _____   30. millimeter _____

25. meter _____   28. milligram _____   31. kiloliter _____

Fill in each blank with the correct equivalent measurement.

32. 1 kg = _____ g   34. 1 m = _____ mm   36. 1 kL = _____ L

33. 1 km = _____ m   35. 1 g = _____ kg   37. 1 L = _____ mL

Name two things that are sold, measured, or packaged in the given unit.

38. meters   _____

39. kilometers   _____

40. grams   _____

41. liters   _____

Reading Rulers and Straight-Line Scales

A ruler is an example of a simple scale. Here are the steps involved in reading a measurement from a ruler.

⎯ Example 1 ⎯

How long is the pencil?

Step 1. Place the ruler so the zero mark on the ruler lines up with one end of the pencil.

Step 2. The units on the ruler are inches. Between which two whole inches is the length of the pencil?

3 and 4

Step 3. Count the number of marks between each inch mark. What does each mark represent?

There are 16 marks between each pair of inch marks. Each mark represents $\frac{1}{16}$ inch.

Step 4. How many $\frac{1}{16}$-inch marks does the pencil extend beyond 3 inches?

Three of them.

Answer: The pencil is about $3\frac{3}{16}$ inches long.

You can read thermometers and many other straight-line scales the same way you read the scale on a ruler.

⎯ Example 2 ⎯

Mark this outdoor thermometer so it shows 73°F. ("°F" stands for "degrees Fahrenheit.")

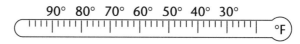

Step 1. Count the number of marks between 70° and 80°. What does each mark represent?

There are 5 marks between 70 and 80. Each mark represents 2°F.

Step 2. How many marks beyond 70° are needed to show 73°F?

$1\frac{1}{2}$ marks

Answer:

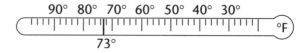

Write the measurement shown on each Fahrenheit or Celsius scale.

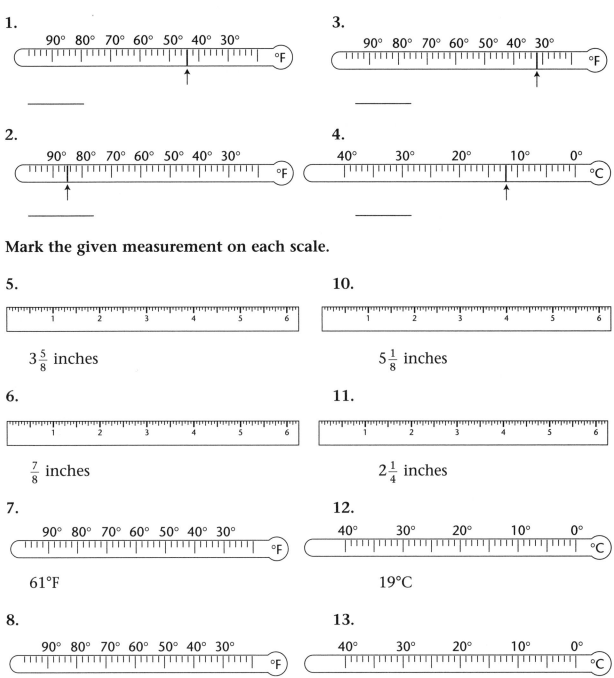

1.

90° 80° 70° 60° 50° 40° 30° °F

_____

2.

90° 80° 70° 60° 50° 40° 30° °F

_____

3.

90° 80° 70° 60° 50° 40° 30° °F

_____

4.

40°   30°   20°   10°   0° °C

_____

Mark the given measurement on each scale.

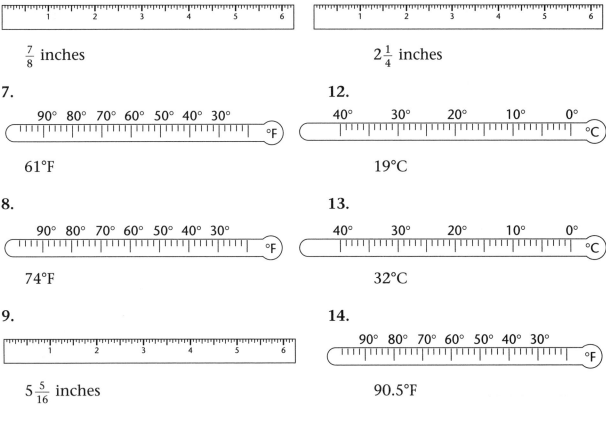

5.

1   2   3   4   5   6

$3\frac{5}{8}$ inches

6.

1   2   3   4   5   6

$\frac{7}{8}$ inches

7.

90° 80° 70° 60° 50° 40° 30° °F

61°F

8.

90° 80° 70° 60° 50° 40° 30° °F

74°F

9.

1   2   3   4   5   6

$5\frac{5}{16}$ inches

10.

1   2   3   4   5   6

$5\frac{1}{8}$ inches

11.

1   2   3   4   5   6

$2\frac{1}{4}$ inches

12.

40°   30°   20°   10°   0° °C

19°C

13.

40°   30°   20°   10°   0° °C

32°C

14.

90° 80° 70° 60° 50° 40° 30° °F

90.5°F

Reading Arcs and Dials

If you have a gas, water, or electric meter at your home, it may have circular dials. Here is an example of how to read a four-dial meter. (Notice that the word *meter*, as used here, is not a unit of length. Its meaning is close to that of a "parking meter.")

— Example 1

Electricity is usually measured in kilowatt-hours (kWh). What is the reading for this four-dial electric meter?

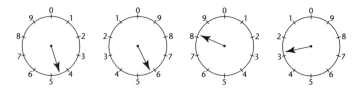

Step 1.	Start at the left. The pointer is between which two whole numbers?	4 and 5
Step 2.	Move to the next dial. The pointer is between which two whole numbers?	5 and 6
Step 3.	For each of the next two dials, the pointer is between which two whole numbers?	8 and 9, 2 and 3

Answer: The reading of the four-dial meter is 4,583 kWh. (The meter-reader usually rounds up for the last dial.)

Express each four-meter reading in kilowatt-hours.

1. 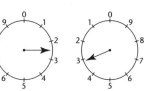   Reading: _____

2. 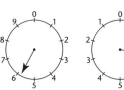   Reading: _____

3. 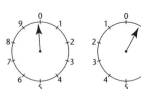   Reading: _____

Example 2

Mark a two-dial meter so it shows a reading of 84 kWh.

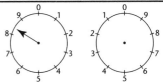

Step 1. Start at the left. Find 8 and 9 on the dial. Draw a pointer between them.

Step 2. Now go to the dial at the right. Find 3 and 4 on the dial. Draw a pointer between them.

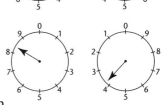

Answer: The two-dial meter shows a reading of 84 kWh.

Mark each four-meter set of dials with the given reading.

4. 2,735 kWh

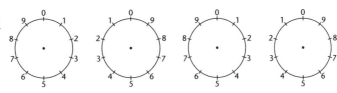

5. 4,508 kWh

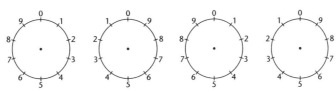

6. 2,995 kWh

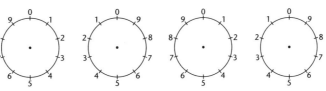

7. 3,009 kWh

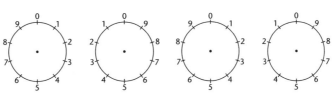

Mark each five-meter set of dials with the given reading.

8. 42,351 kWh

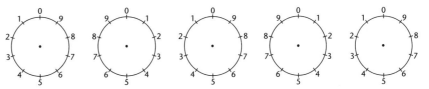

9. 30,037 kWh

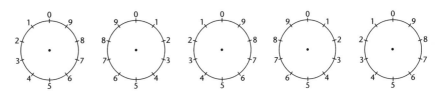

Measuring Around the House

A carpenter's level may use a protractor and a glass tube with an air bubble and water to check the walls and floors of a building. If a wall or the side of a door is straight up-and-down, or vertical, it is called **plumb.** If a floor, ceiling, or doortop is perfectly horizontal, it is called **level.** The protractor is used to measure the angle formed by the wall or floor.

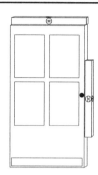

1. Use a carpenter's level and a protractor to describe the angle formed by the walls, floors, and doors of your own house. Are the walls plumb? Are the floors level? If not, what angles do they form with a horizontal line? Many houses "settle" after a few years, when the foundation compresses the earth below it, or the wood in the framing of the house shrinks slightly.

2. Check the bottom of a bathtub, a driveway, or the floor in a basement near a floor drain. What angles do these surfaces form with a horizontal line? Should these surfaces be level?

3. A horizontal line meets a vertical line at a 90-degree angle, or a right angle. Look around your house. Where do you see right angles? Make a list of the 90-degree angles in your house.

The weight of an object is how heavy or light it is. Here are some activities involving weight.

4. Using a bathroom scale, weigh yourself with an overcoat and all your clothes on. Then weigh yourself wearing fewer clothes. How can you find the weight of the clothing you removed without putting that pile of clothes on the scale? What is the weight of the clothing?

5. Use a bathroom scale to weigh a pet or a small child: Weigh yourself holding the pet or child, then weigh yourself alone. What math operation can you use to find the weight of the pet or child? What is that weight?

6. If you have a small scale, weigh some of your letters. How much do they weigh? For a first-class letter, what is the most it can weigh without needing more than a first-class stamp?

7. Before you cook a piece of meat, poultry, or fish, use a kitchen scale to weigh it. Then weigh the food after you cook it. Are the two weights the same? If the weights are different, what do you think happened?

The temperature of a person or an object is a measure of how much heat it contains or is giving off. Here are some activities involving temperature.

8. The "normal" body temperature for people is 98.6°F. This means that most people, when they are healthy, have a temperature close to 98.6°F. Use a personal thermometer to take your temperature before and after exercising. If the two temperatures are different, what do you think caused the difference?

9. Using a thermometer that has a range from at least 40°F to 200°F, find the temperature of a "hot shower," a "cold shower," a "hot bath," and a "cool bath." What is the most comfortable water temperature for you when you shower or bathe? (CAUTION: Do not use a "fever" thermometer for this kind of activity. A temperature above 106°F may destroy that kind of thermometer.)

10. A thermostat helps regulate the temperature of your home. Use a thermometer to find the temperature of several rooms in your home, including places high in the house, low in the house, and far from the thermostat. What is the range in temperature readings? Are they close to the thermostat reading?

11. In your city or town, what is a comfortable setting for a thermostat from December to March? What is a comfortable setting for a thermostat from July to September? If these two settings are different, give an explanation why this may be so.

The perimeter of a figure is the distance around the figure. The area of a figure is the size of the region enclosed by the figure. Here are some activities involving perimeter and area.

12. The figure below at the left shows the floor of a large room. The floor is covered in one-foot by one-foot square tiles. What is the area of the floor? What is the perimeter of the floor? How are the area and perimeter related to the length and width of the floor?

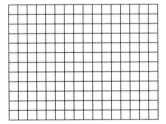   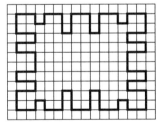

13. The figure at the right above shows the same room with an unusual rug covering part of the floor. What is the area of the rug? What is the perimeter of the rug?

14. In the two figures above, the rug covers part of the floor, so the rug is not as large as the floor. Think about your measurements in Exercises 12 and 13. Can part of the floor ever have a greater area than the whole floor? Can part of the floor ever have a greater perimeter than the whole floor?

The Construction Trades

The home construction industry includes the trades of carpentry, plumbing, roofing, and many others. The trades depend on each other; for example, the drywallers' work is made easier if the carpenters' work is careful and accurate. For all the work to be done properly, all the workers must use careful and accurate measurements.

Think about all the measurements that go into building a house. A surveyor provides an accurate outline of the land, and an architect makes scale drawings for the builders. Excavators dig out earth, moving it around to be sure water will drain away from the house. Then concrete workers pour a foundation.

Carpenters mark, cut, and nail together the framing, including openings for doors and windows. Roofers shingle the top, and other workers provide siding and install windows. Inside the house, plumbers and electricians run pipes and cables through the walls, as do heating, air-conditioning, and telephone workers. Other workers install insulation, drywall, and kitchen and bathroom fixtures. Finally, painters, tilers, wallpapers, and carpet layers take their turns.

Outside, landscape workers add topsoil, then plant grass seed and ground cover. Finally, the homeowner may build a fence to protect the new grass and buy curtains or blinds for the windows.

How many things can you name that were measured? What units do you think were used? How many measuring tools can you name that may have been used?

Suppose you are the homeowner or general contractor for a new house. Make up answers to these questions, which are just some of the measurement problems you would have to answer about the house.

- **What is the shape of the property, and what is the shape of the house?**

- **What is the size of the front yard and the back yard? How large is the garage?**

- **How many linear feet of lumber are needed for the framing?**

- **How many square feet of shingles are needed for the roof? How much siding is needed? How much trim for the windows?**

- **How many gallons of paint are needed? How many square feet of tile, carpet, and other flooring?**

> As you can tell, many of the problems in designing and building a house involve length, angles, and area. In these lessons, you will learn about measuring length, measuring angles, and calculating area. Also, you will investigate tools for measuring length, angles, and area, and apply those tools to solving problems.

Career Corner

Here are some of the jobs in the construction field that were mentioned.

air-conditioning worker	electrician	plumber
architect	excavator	roofer
carpenter	heating contractor	surveyor
carpet layer	insulator	telephone installer
concrete worker	landscape contractor	tiler
drywall installer	painter	wallpaperer

There are many other jobs in the construction field, including glaziers, bricklayers, garage door installers, and floor sanders. Can you think of any other jobs to add to the list?

Estimating Length in Meters

The basic metric units for length are **millimeter**, **centimeter**, **meter**, and **kilometer**. Here are the relationships among these units.

1 meter (m)	=	100 centimeters (cm)	1 cm = 10 mm
1 m	=	1,000 millimeters (mm)	1 km = 1,000 m

width of a
needle: 1 mm

width of your
little finger: 1 cm

height of a
six-year-old: 1 m

distance around a
large museum: 1 km

Example 1

Janis used a meter stick, which is a ruler that is 1 meter long, to measure the thickness of a telephone directory. What was the thickness of the directory?

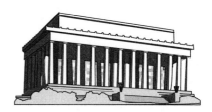

centimeters (not to scale)

Step 1. Place the meter stick so one end is even with one of the covers.

Step 2. The numbers on the meter stick are centimeters. Between which two whole centimeters is the other cover?   11 and 12

Step 3. Read the number on the meter stick.   about 11.9 cm

Answer: The thickness of the directory is about 11.9 cm.

Name three things that might have each given length.

1. 1 millimeter

_____

2. 1 centimeter

_____

3. 1 meter

_____

4. 1 kilometer

_____

Write the measurement shown by each arrow.

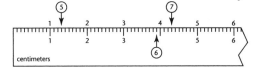

5. _____   6. _____   7. _____

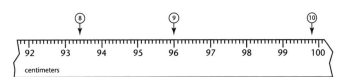

8. _____   9. _____   10. _____

Use an arrow and a number to mark each measurement on the metric ruler.

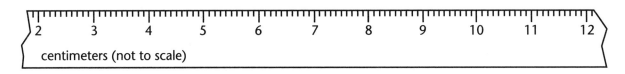

11. 5 cm   12. 8 cm 6 mm   13. 4 cm 8 mm

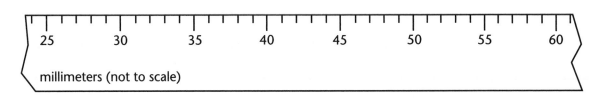

14. 28 mm   15. 32 mm   16. 51 mm

Metric Lengths and the Decimal Point

Metric units use a decimal to represent a part of a unit.

1 m	=	100 cm	so	1 cm	= 0.01 m
1 m	=	1,000 mm	so	1 mm	= 0.001 m
1 km	=	1,000 m	so	1 m	= 0.001 km

Example 1

The widths of some metric machine parts are 19 millimeters, 48 millimeters, and 197 millimeters. How can you express each measurement in centimeters?

Step 1. Write the relationship between mm and cm.

$$10 \text{ mm} = 1 \text{ cm}$$
$$1 \text{ mm} = 0.1 \text{ cm}$$

Step 2. To change from mm to cm, divide by 10. This is the same as moving the decimal point one place to the left.

$$19 \text{ mm} = \frac{19}{10} \text{ cm} = 1.9 \text{ cm}$$
$$48 \text{ mm} = \frac{48}{10} \text{ cm} = 4.8 \text{ cm}$$
$$197 \text{ mm} = \frac{197}{10} \text{ cm} = 19.7 \text{ cm}$$

Answer: The measurements are 1.9 centimeters, 4.8 centimeters, and 19.7 centimeters.

Example 2

Express the measurements from Example 1 in meters.

Step 1. Write the relationship between millimeters and meters.

$$1,000 \text{ mm} = 1 \text{ m}$$
$$1 \text{ mm} = 0.001 \text{ m}$$

Step 2. To change from mm to m, divide by 1,000. This is the same as moving the decimal point three places to the left.

$$19 \text{ mm} = \frac{19}{1,000} \text{ m} = 0.019 \text{ m}$$
$$48 \text{ mm} = \frac{48}{1,000} \text{ m} = 0.048 \text{ m}$$
$$197 \text{ mm} = \frac{197}{1,000} \text{ m} = 0.197 \text{ m}$$

Answer: The measurements are 0.019 meters, 0.048 meters, and 0.197 meters.

Rewrite each measurement so it uses the given unit.

1. 2,000 m = ___ km

2. 1,000,000 cm = ___ m

3. 750,000 cm = ___ km

4. 300 mm = ___ cm

5. 600 m = ___ km

6. 1,000,000 mm = ___ m

7. 2,200 m = ___ km

8. 625 cm = ___ m

9. 1,475 mm = ___ cm

10. 8,700 mm = ___ m

11. 4,900 m = ___ km

12. 18,800 cm = ___ km

Example 3

The distance between two towns is 15.83 kilometers. Explain how to change 15.83 kilometers to meters.

Step 1. Write the relationship between km and m.   1 km = 1,000 m

Step 2. To change km to m, multiply by 1,000. This is the same as moving the decimal point three places to the right.

15.83 km = (15.83)(1,000) m
= 15,830 m

Answer: 15.83 km = 15,830 m

Rewrite each measurement so it uses the given unit.

13. 3 m = ___ cm

14. 529 cm = ___ mm

15. 27 m = ___ cm

16. 21 km = ___ m

17. 7 km = ___ cm

18. 12 m = ___ mm

19. 2.25 m = ___ mm

20. 3.2 cm = ___ mm

21. 40.5 km = ____ m

22. 1.6 km = ___ m

23. 12.75 cm = ___ mm

24. 5.2 km = ___ m

25. 8.2 km = ___ cm

26. 3.6 m = ___ cm

Solve each problem.

27. Matt measured a board and found it was 30.6 centimeters wide. How many millimeters is that?

28. Tim, a delivery person, drove 21.68 kilometers. How many meters did he drive?

29. Jana measured the distance around the picture frame as 180 centimeters. How many meters is that?

30. Georgia said she had just finished jogging 3,500 meters. How many kilometers did she run?

Operations with Metric Measurements

To add, subtract, multiply, or divide with metric measurements, express each measurement using just one unit. Then use paper and pencil or a calculator to do the operation.

Example 1

A length of pipe 3 meters and 15 centimeters long is cut into 4 equal pieces. What is the length of each piece?

Step 1.	Write the measurement using a single unit.	$3 \text{ m} + 15 \text{ cm} = 3\text{m} + \frac{15}{100} \text{ m}$ $= 3 \text{ m} + .15 \text{ m}$ $= 3.15 \text{ m}$
Step 2.	Divide 3.15 by 4.	$3.15 \div 4 = 0.7875$

Answer: The length of each piece of pipe is 0.7875 m. That is the same as 78.75 cm.

Example 2

Add: 3.75 meters plus 850 centimeters.

Step 1.	Write 850 cm in meters.	$850 \text{ cm} = 8.5 \text{ m}$
Step 2.	Add the numbers.	$3.75 + 8.5 = 12.25$

Answer: The total is 12.25 m.

For each operation, use a single unit to find the answer.

1. $15 \times (10 \text{ m} + 350 \text{ cm})$

2. $6 \times (2 \text{ km} + 135 \text{ m})$

3. $(4 \text{ cm} + 18 \text{ mm}) \div 5$

4. $450 \text{ mm} - 2 \text{ cm}$

5. $3 \text{ km} + 300 \text{ m} + 300 \text{ mm}$

6. $(4 \text{ m} + 37 \text{ cm}) + (17 \text{ m} + 250 \text{ cm})$

Solve each problem.

7. The length of a rectangular field is 0.015 kilometers and the width is 25 meters. What is the length of a fence that goes all around the field?

8. To build the fence in Exercise 7, one of the 0.015-kilometer sides uses 5 equal sections of fencing. How long is each section of fencing?

Estimating Length Using Customary Units

The customary units of length probably are familiar to you: **inches**, **feet**, **yards**, and **miles**. Here are the relationships among these units.

12 inches (in.)	=	1 foot (ft)	5,280 ft	= 1 mile (mi)
3 ft	=	1 yard (yd)	1,760 yd	= 1 mi
36 in.	=	1 yd		

height of a
paper clip: 1 in.

height of this
book: just under 1 ft

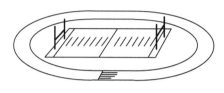

four times around
this track: 1 mi

Name four things that might be measured in each unit.

1. inches

_____

2. feet

_____

3. yards

_____

4. miles

_____

Circle the item you think is longer.

5. a pencil or a jump rope

6. a hammer or a nail file

7. an extension cord or a yardstick

8. length of a key or a screwdriver

Circle the more appropriate measurement for the length of each object.

9. height of a 6-year-old child
   38 in. or 38 ft

10. distance of a "sprint" race
    100 yd or 100 mi

11. length of a window frame
    42 ft or 42 in.

12. length of a bed sheet
    86 yd or 86 in.

13. height of a giraffe
    18 yd or 18 ft

14. height of a flag pole
    60 ft or 60 yd

15. distance from Chicago to St. Louis
    325 mi or 325 ft

16. length of a garden hose
    50 ft or 50 yd

17. length of a swimming pool
    25 yd or 25 in.

18. distance from Portland, OR, to L.A.
    410 yd or 410 mi

In each group, underline the shortest object.

19. a pair of pliers
    a pair of tongs
    a spatula

20. a man
    a child
    a pony

21. a postcard
    a business envelope
    a roll of adhesive tape

22. a roll of wallpaper
    a roll of toilet tissue
    a roll of aluminum foil

23. a chair
    a curtain for a patio door
    a dish towel

Example

Jason wanted to use a yardstick to measure the distance between two pencil marks on a board. A yardstick is a ruler that is 36 inches, or 1 yard, in length. What is the distance between the two pencil marks?

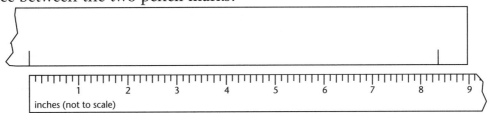

Step 1. Place the yardstick so one of the pencil marks is at the zero point.

Step 2. Between which two whole inches is the second pencil mark?

8 and 9

Step 3. Look at the number of tick marks on the yardstick between 8 and 9. What does each tick mark represent?

There are 8 tick marks, so each tick mark represents $\frac{1}{8}$ inch.

Step 4. Read the distance between the pencil marks.

8 inches and $\frac{3}{8}$ of an inch

Answer: The distance between the two pencil marks is $8\frac{3}{8}$ inches.

Write the measurement shown by each arrow.

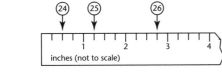

24. _____   **25.** _____   **26.** _____

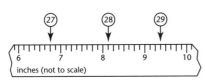

27. _____   **28.** _____   **29.** _____

Label the yardstick to show each length.

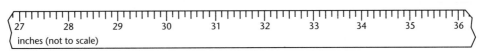

30. $31\frac{5}{8}$ in.   **32.** $32\frac{1}{2}$ in.   **34.** $35\frac{7}{8}$ in.

31. 34 in.   **33.** $33\frac{1}{4}$ in.   **35.** $35\frac{1}{8}$ in.

Changing Among Inches, Feet, Yards, and Miles

What happens to the measurement of an object when you use a different unit? Here are the relationships between inches, feet, yards, and miles.

12 inches (in.)	=	1 foot (ft)
3 ft	=	1 yard (yd)
36 in.	=	1 yd
5,280 ft	=	1 mile (mi)
1,760 yd	=	1 mi

Example 1

Tom wants to put a fence around his garden. He measured the distance around the garden and found it to be 24 yards. The fencing is sold in feet. How many feet will he need?

Step 1. Write the problem.   24 yd = ____ ft

Step 2. Write the relationship between yards and feet.   1 yd = 3 ft

Step 3. Multiply the number of yards by 3. $3 \times 24 = 72$

Answer: 24 yd = 72 feet. Tom needs 72 feet of fencing for his garden.

Example 2

Change 96 inches to feet.

Step 1. Write the problem.   96 in. = ___ ft

Step 2. Write the relationship between feet and inches.   12 in. = 1 ft

Step 3. We want to go from 12 inches to 96 inches, and $12 \times 8 = 96$. So multiply each side of the equation in Step 2 by 8.   96 in. = 8 ft

Answer: 96 in. = 8 ft

> In general, when you change from a smaller unit to a larger one, the number of units in your measurement will decrease. When you change from a larger unit to a smaller one, the number of units in your measurement will increase.

Rewrite each measurement so it uses the given unit.

1. 9 ft = ___ in.

2. 2 yd = ___ in.

3. 10 yd = ___ ft

4. 30 ft = ___ in.

5. 4 mi = ___ ft

6. 20 mi = ___ yd

7. 28 yd = ___ ft

8. 10 mi = ___ ft

9. 3 mi = ___ yd

Rewrite each measurement so it uses the given unit.

10. $5\frac{1}{2}$ ft = _____ in.

11. $6\frac{1}{2}$ yd = _____ ft

12. $10\frac{1}{4}$ mi = _____ yd

13. 24 in. = _____ ft

14. 72 in. = _____ yd

15. 33 ft = _____ yd

16. 14 ft = _____ yd

17. 242 ft = _____ yd

18. 6 in. = _____ ft

19. $2\frac{1}{10}$ mi = _____ ft

20. 8.5 ft = _____ in.

21. 7.2 yd = _____ in.

22. 108 in. = _____ yd

23. 72 in. = _____ ft

24. 6 ft = _____ yd

25. 26 in. = _____ ft

26. 30 in. = _____ ft

27. 48 in. = _____ yd

28. $1\frac{1}{4}$ mi = _____ ft

29. $12\frac{1}{3}$ yd = _____ ft

30. $2\frac{1}{6}$ yd = _____ in.

31. 10,560 ft = _____ mi

32. 240 in. = _____ ft

33. 26,400 ft = _____ mi

34. 7 ft = _____ yd

35. 2,640 yd = _____ mi

36. 78 in. = _____ yd

Solve each problem.

37. Keisha bought $2\frac{1}{3}$ yards of wood framing for a picture. How many feet did she buy?

38. Eric drove the delivery truck 33 miles on Saturday. How many feet did he drive?

39. Toni runs for exercise. Yesterday her route was $1\frac{1}{2}$ miles long. How many yards did she run?

40. Lee plans to knit an afghan throw for her mother. She needs 80 yards of yarn. How many feet of yarn will she have to buy?

41. George decided to plant shrubs 1 foot apart across the front of his house. His yard is 66 yards wide. How many shrubs will he need?

42. Juan wants a new TV for his family room. The TV cabinet space is two and a half feet wide. What is the width, in inches, of the largest cabinet he can buy?

43. Elsa was asked to put lace around the bottom of a ballerina's skirt. The length around the skirt is 94 inches. The lace comes in 6-foot packages. Will one package be enough for Elsa to use?

Adding and Subtracting Lengths

Sometimes you have to add or subtract lengths that contain different units. One way to add and subtract these measurements is to regroup.

—— Example 1 ——

Jessie bought a copper pipe that was 7 feet 10 inches long and another one that was 4 feet 5 inches long. How much copper pipe did Jessie buy?

Step 1.	Line up the measurements, putting like units under like units.	7 ft 10 in. + 4 ft 5 in.
Step 2.	Add the inches and add the feet.	7 ft 10 in. + 4 ft 5 in. 11 ft 15 in.
Step 3.	Change the inches to feet and inches.	15 in. = 1 ft 3 in.
Step 4.	Rewrite the total.	11 ft + 15 in. 11 ft + 1 ft + 3 in. 12 ft 3 in.

Answer: Jessie bought 12 feet 3 inches of copper pipe.

Find each sum.

1. 3 yd 2 ft + 6 yd 2 ft

2. 7 ft 10 in. + 2 ft 5 in.

3. 4 yd 21 in. + 3 yd 18 in.

4. 6 ft 11 in. + 2 ft 7 in.

5. 2 yd 8 in. + 3 yd 10 in.

6. 4 yd 28 in. + 2 yd 16 in.

Sometimes you need to add measurements that have two or more units. To add these measurements, the first step is to line up the units that are the same.

Example 2

Add 2 ft 8 in. and 2 yd 21 in.

Step 1.	Line up the measurements, putting like units under like units.	2 ft 8 in. + 2 yd 21 in.
Step 2.	Add the like units.	2 ft 8 in. + 2 yd 21 in. 2 yd 2 ft 29 in.
Step 3.	Change the inches to feet and inches.	29 in. = 24 in. + 5 in. = 2 ft 5 in.
Step 4.	Rewrite the total.	2 yd + 2 ft + 29 in. 2 yd + 2 ft + 2 ft + 5 in. 2 yd + 4 ft + 5 in. 3 yd + 1 ft + 5 in.

Answer: The total is 3 yd 1 ft 5 in.

Example 3

Subtract 1 ft 8 in. from 3 ft 4 in.

Step 1.	Line up the measurements, putting like units under like units.	3 ft 4 in. – 1 ft 8 in.
Step 2.	You have to borrow. The four lines at the right show that 3 ft 4 in. is the same as 2 ft 16 in.	3 ft + 4 in. 2 ft + 1 ft + 4 in. 2 ft + 12 in. + 4 in. 2 ft 16 in.
Step 3.	Rewrite the problem and subtract.	2 ft 16 in. – 1 ft 8 in. 1 ft 8 in.

Answer: The difference is 1 ft 8 in.

Example 4

Add 3 yd 28 in. + 6 yd 22 in.

| Step 1. | Line up the units. Use "0 ft" for the number of feet. | 3 yd 0 ft 28 in.
 + 6 yd 0 ft 22 in. |

| Step 2. | Add the yards, feet, and inches. | 3 yd 0 ft 28 in.
 + 6 yd 0 ft 22 in.
 9 yd 0 ft 50 in. |

| Step 3. | Change the inches to yards, feet, and inches. | 50 in. = 36 in. + 12 in. + 2 in.
 = 1 yd 1 ft 2 in. |

| Step 4. | Rewrite the total. | 9 yd + 0 ft + 50 in.
 9 yd + 1 yd + 1 ft + 2 in.
 10 yd 1 ft 2 in. |

Answer: The sum is 10 yd 1 ft 2 in.

Add or subtract.

7. 2 yd 6 ft + 1 ft 12 in.

8. 4 ft 3 in. − 2 ft 7 in.

9. 3 ft 4 in. − 1 ft 11 in.

10. 1 yd 1 ft 11 in. + 3 yd 2 ft 8 in.

11. 4 yd 1 ft − 1 yd 2 ft

12. 6 yd 1 ft 8 in. + 2 yd 2 ft 3 in.

13. 3 ft 8 in. + 2 yd 29 in.

14. 16 yd − 8 yd 2 ft

15. 6 ft 7 in. − 2 ft 9 in.

16. 2 yd 2 ft 14 in. + 3 yd 26 in.

17. 4 ft − 2 ft 5 in.

18. 6 yd 2 ft + 3 ft 10 in.

Solve each problem.

19. Carlos cut a piece of wood 3 feet 7 inches long from a board that was 7 feet 3 inches long. How long is the piece that he has left?

20. Joyce measured one wall at 12 feet 7 inches. Another wall is 11 feet 11 inches. What is the total height of the two walls?

Multiplying and Dividing Lengths

To multiply or divide with lengths, you use the same two steps but in the opposite order. To multiply a length times a number, first perform the operation and then change the units. To divide a length by a number, first change the units and then perform the operation.

— Example 1

Rafael wanted to make 4 shelves that each were 3 feet 8 inches long. How many feet of board will he have to buy?

Step 1. Decide which math operation to use. The operation to use is multiplication.

Step 2. Write the problem.

$$\begin{array}{r} 3 \text{ ft} \quad 8 \text{ in.} \\ \times \qquad 4 \\ \hline \end{array}$$

Step 3. Multiply the inches and the feet by 4.

$$\begin{array}{r} 3 \text{ ft} \quad 8 \text{ in.} \\ \times \qquad 4 \\ \hline 12 \text{ ft } 32 \text{ in.} \end{array}$$

Step 4. Change the inches to feet and inches.

$$\begin{aligned} 32 \text{ in.} &= 24 \text{ in.} + 8 \text{ in.} \\ &= 2 \text{ ft} \quad 8 \text{ in.} \end{aligned}$$

Step 5. Rewrite the product.

12 ft + 32 in.
12 ft + 24 in. + 8 in.
14 ft 8 in.

Answer: Rafael must buy a board that is at least 14 feet 8 inches long.

— Example 2

Divide 9 yards 1 foot by 4.

Step 1. Write the problem.

$$\frac{9 \text{ yd } 1 \text{ ft}}{4} = ?$$

Step 2. Change the numerator to feet.

9 yd 1 ft = 27 ft + 1 ft = 28 ft

Step 3. Rewrite the problem.

$$\frac{9 \text{ yd } 1 \text{ ft}}{4} = \frac{28 \text{ ft}}{4} = ?$$

Step 4. Divide. Then change back to yards and feet.

$$\begin{aligned} \frac{28 \text{ ft}}{4} &= 7 \text{ ft} \\ &= 6 \text{ ft} + 1 \text{ ft} \\ &= 2 \text{ yd } 1 \text{ ft} \end{aligned}$$

Answer: The quotient is 2 yards 1 foot.

Multiply or divide.

1. 6 ft 4 in. × 5 =

2. 6 yd 2 ft ÷ 2 =

3. 10 ft 3 in. ÷ 3 =

4. 12 yd ÷ 2 =

5. 5 yd 1 ft ÷ 4 =

6. 8 in. ÷ 2 =

7. 9 yd 1 ft ÷ 4 =

8. 3 ft 10 in. ÷ 2 =

9. 7 ft 6 in. ÷ 6 =

10. 5 ft 4 in. × 6 =

11. 3 ft 6 in. × 5 =

12. 16 ft 4 in. ÷ 7 =

13. 15 yd ÷ 5 =

14. 10 ft 5 in. ÷ 5 =

15. 6 yd 2 ft × 8 =

16. 4 yd 10 in. × 5 =

17. 6 ft × 5 =

18. 2 ft 4 in. × 8 =

19. 6 ft 4 in. ÷ 4 =

20. 4 ft 9 in. ÷ 3 =

21. 3 ft 11 in. × 0 =

Solve each problem.

22. Jen has a 10 foot 4 inch board to make shelves. If she wants 4 shelves of equal lengths, how long will each shelf be?

23. Mario needs 2 yards 8 inches of fabric to cover a table. If he wants to cover 4 tables, how much material does he need?

24. Camelia was making oak wood picture frames. She measured the length around one frame as 40 inches. If she wants to make five frames, how much oak does she need? (Answer in feet and inches.)

25. Brian wants to divide a package of insulation into three equal amounts. If the package says $2\frac{1}{2}$ yards, how many feet and inches will each piece be?

Comparing and Ordering Customary and Metric Units of Length

Is a kilometer longer or shorter than a mile? Is an inch longer or shorter than a centimeter? Here are some approximate comparisons that will help you compare the customary units of inches, feet, and miles to the metric units of centimeters, meters, and kilometers.

1 inch	≈ 2.54 centimeters	1 centimeter	≈	$\frac{3}{8}$ inch
1 foot	≈ 30 centimeters	1 meter	≈	39.37 inches
1 mile	≈ 1.6 kilometers	1 kilometer	≈	0.6 mile

Misty wants to run a 10-kilometer marathon race, but she is not sure if she can run that far. To find out, she can approximate 10 kilometers with some number of miles.

Step 1. Write the problem.   10 km ≈ ____ miles

Step 2. Write the relationship   1 km ≈ 0.6 miles
between kilometers and miles.

Step 3. We want to go from 1 km to 10 km. So   10 km ≈ 6 miles
multiply both sides of the equation
in Step 2 by 10.

Answer: If Misty can run about 6 miles, then she can finish a 10-kilometer race.

Circle the larger measurement.

1. 1 in. or 1 cm   4. 1 ft or 1 m   7. 1 mi or 1 km

2. 1 m or 1 km   5. 1 cm or 1 ft   8. 1 ft or 1 km

3. 1 cm or 1 yd   6. 1 km or 1 in.   9. 1 m or 1 yd

Circle the more appropriate measurement for the length of each object.

10. length of a swimming pool   14. width of a business envelope
    25 m or 25 in.   10 cm or 10 m

11. length of a telephone cord   15. length of a pencil
    24 yd or 2 cm   15 in. or 15 cm

12. length of a beach towel   16. size of a 20-year-old's waist
    60 in. or 60 mm   28 in. or 28 mm

13. length of an umbrella   17. length of a queen-size sheet
    1 yd or 1 cm   3 yd or 3 ft

Comparing and Ordering Lengths

To find out which of two lengths is longer or shorter than the other, it is easier if you change the measurements to the same unit.

Example 1

Miranda wants to put insulation around a small window. She measured the distance around the window as 130 inches. The insulation comes in packages that contain 3 yards of material. How many packages does she need?

Step 1.	Write the two measurements.	3 yards, 130 inches
Step 2.	Write the relationship between inches and yards.	1 yd = 36 in.
Step 3.	Decide whether to change inches to yards or yards to inches. Then tell how to change from one unit to the other.	Change yards to inches. Multiply the number of yards by 36.
Step 4.	Change yards to inches.	3 • 36 = 108, so 3 yd = 108 in.
Answer:	One package contains 3 yards, or 108 inches, of insulation. Miranda needs 130 inches of insulation, so she should buy 2 packages.	

Example 2

Which is shorter, 2.5 meters or 220 centimeters?

Step 1.	Write the two measurements.	2.5 m, 220 cm
Step 2.	Write the relationship between meters and centimeters.	1 m = 100 cm
Step 3.	Decide whether to change meters to centimeters or centimeters to meters. Then tell how to change from one unit to the other.	Change meters to centimeters. Multiply the number of meters by 100.
Step 4.	Change meters to centimeters.	2.5•100 = 250, so 2.5 m = 250 cm
Step 5.	Compare and order the measurements.	220 < 250
Answer:	220 centimeters is shorter than 2.5 meters.	

Circle the smaller measurement.

1. 35 cm or 3 mm

2. 6 ft or 60 in.

3. 65 m or 6 km

4. 2 yd or 7 ft

5. 2,400 m or 3 km

6. 48 in. or 5 ft

7. $6\frac{1}{2}$ m or 570 cm

8. 2 mi or 10,000 ft

9. 3 m or 350 cm

Fill in each blank with <, >, or = to represent "is less than," "is greater than," or "is equal to."

10. 21 ft ____ 13 yd

11. 11 mi ____ 15,000 yd

12. 2 m ____ 200 cm

13. 60 mm ____ 6 cm

14. 1.5 m ____ 7,500 mm

15. 36 in. ____ 3 ft

16. 4 yd ____ 60 in.

17. 2,800 m ____ 3 km

18. 21,000 yd ____ 5 mi

19. 2 in. ____ $\frac{1}{2}$ ft

Write the measurements in order from longest to shortest.

20. 2 in., 2 yd, 2 ft, 2 mi

_____

21. 6 cm, 67 mm, 35 cm, 3 m

_____

22. 26 yd, 18 ft, 4 yd, 3 mi

_____

23. 20 mm, 4 m, 350 cm, 6 cm

_____

Solve each problem.

24. Kai bought 3 yards of oak floorboard. Is that more or less than 10 feet?

25. Molly is a nurse. She measured a child as 3 feet tall. Is the little girl more than, less than, or equal to a yard tall?

26. Joan wanted to sew trim around a tablecloth. The distance around the cloth is 260 inches. That trim comes in a 4-yard package. Is one package enough trim to go around the tablecloth?

27. A sign at the amusement park said a person must be taller than 52 inches to go on the ride. Jessica is 4 feet tall. Can she go on the ride?

Focus on Geometry: Measuring Angles

An **angle** is formed when two lines start at the same point. That point is called the **vertex** of the angle, and the lines are called the **sides** of the angle.

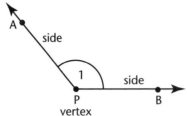

Using the symbol ∠ for angle, you can name an angle three ways:

1. ∠APB   You can name an angle with three letters. The vertex is always the middle letter.
2. ∠P   If a point is the vertex for only one angle, you can name the angle using that single letter.
3. ∠1   You can name an angle by writing a number or letter inside the angle.

Write other names for each angle.

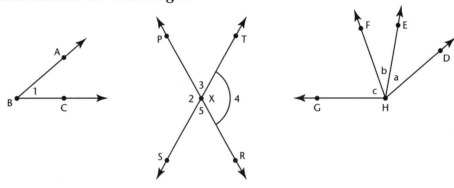

1. ∠1   3. ∠3   5. ∠5   7. ∠b

2. ∠2   4. ∠4   6. ∠a   8. ∠c

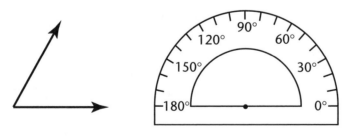

Angles are measured in **degrees**. The tool used to measure an angle is called a **protractor**. A protractor shows measurements along a semicircle from 0° to 180°. It has a dot for the vertex of the angle.

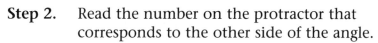

Example 1

To use a protractor to measure an angle, follow these steps.

Step 1. Place the protractor. The vertex of the angle should be under the center of the semicircle. One of the sides of the angle should go through the 0° mark on the protractor.

Step 2. Read the number on the protractor that corresponds to the other side of the angle.

Answer: The measurement of the angle is 60°.

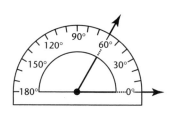

The number on the protractor is 60°.

Example 2

Using a ruler and a protractor, draw an angle of 70°.

Step 1. Draw the vertex and one side of the angle.

Step 2. Place the protractor. The vertex of the angle should be under the center of the semicircle. The side of the angle should go through the 0° mark.

Step 3. Make a pencil mark where the protractor shows 70°.

Answer: Draw the other side of the angle and label the angle with its measurement, which is 70°.

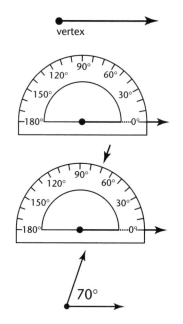

Use a protractor to measure each angle.

9. 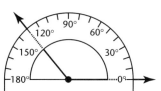   10. 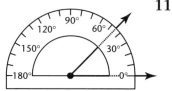   11.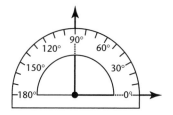

Use a protractor to draw each angle.

12. 120°   14. 25°   16. 180°

13. 115°   15. 90°   17. 65°

Focus on Geometry: Kinds of Angles

There are four different kinds of angles: acute, right, obtuse, and straight.

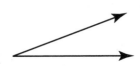

An **acute angle** is an
angle between 0° and 90°.

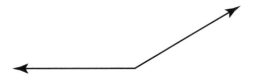

An **obtuse angle** is an
angle between 90° and 180°.

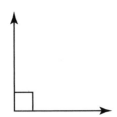

A **right angle** is an angle
that is exactly 90°. A right
angle can be shown
with a square corner at
the vertex.

A **straight angle** is an angle
that is exactly 180°. A straight
angle looks like a line.

Label each angle as acute, right, obtuse, or straight. Then use a protractor to measure the angle.

1.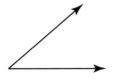

2.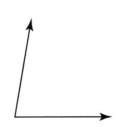

3.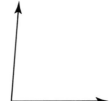

4.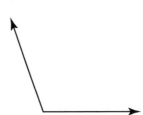

5.

6.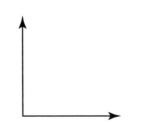

Draw each type of angle. Then use a protractor to measure the angle.

7. an acute angle

8. an obtuse angle

9. a straight angle

10. a right angle

Focus on Geometry: Angles and Parallel Lines

Two lines that are always the same distance apart, and never meet, are called **parallel lines**. The symbol // indicates lines that are parallel, so the expression $\overleftrightarrow{AB}$ // $\overleftrightarrow{CD}$ is read, "line AB is parallel to line CD."

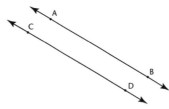

Example 1

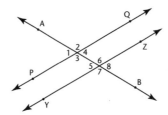

Lines $\overleftrightarrow{PQ}$ and $\overleftrightarrow{YZ}$ are parallel. Which angles are equal to each other?

Step 1. Use a protractor to measure all the angles.

∠1 = 60°, ∠2 = 120°
∠3 = 120°, ∠4 = 60°
∠5 = 60°, ∠6 = 120°
∠7 = 120°, ∠8 = 60°

Step 2. List the angles that are equal to each other.

Answer: ∠1 = ∠4 = ∠5 = ∠8, ∠2 = ∠3 = ∠6 = ∠7

Some pairs of angles are called **corresponding angles**. Examples are ∠1 and ∠5, ∠3 and ∠7, ∠2 and ∠6, and ∠4 and ∠8. When two parallel lines are cut by a third line, the corresponding angles are always equal.

Some pairs of angles are called **alternate interior angles**. There are two pairs: ∠3 and ∠6, and ∠4 and ∠5. When two parallel lines are cut by a third line, the alternate interior angles are always equal.

The two horizontal lines at the right are parallel. Find the measure of each angle without using a protractor.

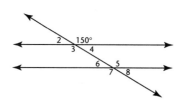

1. ∠2 = _____   4. ∠6 = _____   7. ∠5 = _____

2. ∠3 = _____   5. ∠7 = _____   8. ∠5 + ∠6 = _____

3. ∠4 = _____   6. ∠8 = _____   9. ∠3 + ∠4 = _____

Focus on Geometry: Pairs of Angles

When two angles form a right angle or add up to 90°, they are called **complementary angles.** When two angles form a straight angle or add up to 180°, they are called **supplementary angles.**

— Example 1 —

Josie was fitting two pieces of glass together to form a right angle for a stained-glass window. If one piece of glass is cut at an angle of 30°, what is the angle for the other piece of glass?

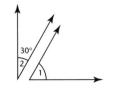

Step 1.   What is the sum of the two angles? $\angle 1 + \angle 2 = 90°$

Step 2.   Substitute 30° for $\angle 2$. $\angle 1 + 30° = 90°$

Step 3.   Subtract 30° from each side. $\angle 1 = 60°$

Answer: The angle for the other piece of glass is 60°.

Find the complement of each angle.

1. 75° _____   3. 68° _____   5. 39° _____

2. 31° _____   4. 47° _____   6. 86° _____

Find the complement of each angle.

7.   8.

complement: _____   complement: _____

Example 2

Find the supplement of an angle
whose measure is 50°.

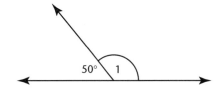

Step 1. Write the sum of the two angles. $\angle 1 + 50° = 180°$

Step 2. Subtract 50° from each side. $\angle 1 = 130°$

Answer: The measurement of the supplement is 130°.

Find the supplement of each angle.

9. 80° _____   11. 40° _____   13. 145° _____

10. 120° _____   12. 25° _____   14. 68° _____

Find the supplement of each angle. Describe the supplement as acute, right, obtuse, or straight.

15.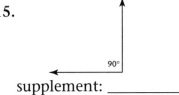

supplement: _____

type: _____

18.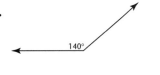

supplement: _____

type: _____

16.

supplement: _____

type: _____

19.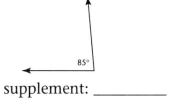

supplement: _____

type: _____

17.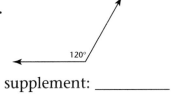

supplement: _____

type: _____

20.

supplement: _____

type: _____

Focus on Geometry: Area and Perimeter

The **perimeter** of a figure is the distance around it.
The **area** of a figure is the size of the region it covers.
Area is measured in square units.

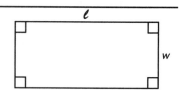

For a rectangle with length ℓ and width w, formulas for
the perimeter P and the area A are:

$$P = 2\ell + 2w \qquad A = \ell \cdot w$$

Example 1

Carla's living room is 10 feet by 12 feet. How much wood trim does she need for the
perimeter of the room? How much carpet does she need for the area of the room?

Step 1.	Write the dimensions of the room.	$\ell = 12$ ft, $w = 10$ ft
Step 2.	To find the perimeter, substitute the values of ℓ and w into the formula $P = 2\ell + 2w$.	$P = 2\ell + 2w$ $P = 2(12) + 2(10)$ $= 24 + 20 = 44$ ft
Step 3.	To find the area, substitute the values of ℓ and w into the formula $A = \ell \cdot w$.	$A = \ell \cdot w$ $= (12)(10)$ $= 120$ sq ft

Answer: The perimeter is 44 feet and the area is 120 square feet.

Example 2

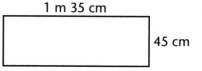

Find the perimeter and the area of this rectangular cloth.

Step 1.	Start by using a single unit for the length and the width.	Change to centimeters: $w = 45$ cm $\ell = 1$ m $+ 35$ cm $= 100$ cm $+ 35$ cm $= 135$ cm
Step 2.	To find the perimeter, substitute the values for ℓ and w into the formula $P = 2\ell + 2w$.	$P = 2\ell + 2w$ $P = 2(135) + 2(45)$ $= 270 + 90 = 360$ cm
Step 3.	To find the area, substitute the values for ℓ and w into the formula $A = \ell \cdot w$.	$A = \ell \cdot w$ $= (135)(45)$ $= 6,075$ sq cm

Answer: The perimeter of the cloth is 360 centimeters and its area is
6,075 square centimeters.

Find the perimeter and area of each figure.

1.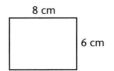

8 cm

6 cm

P = _____

A = _____

4.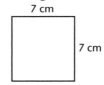

7 cm

7 cm

P = _____

A = _____

7.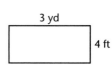

3 yd

4 ft

P = _____

A = _____

2.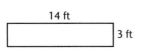

14 ft

3 ft

P = _____

A = _____

5.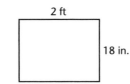

2 ft

18 in.

P = _____

A = _____

8.

20 cm

10 mm

P = _____

A = _____

3.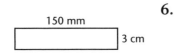

150 mm

3 cm

P = _____

A = _____

6.

13.5 cm

12.2 cm

P = _____

A = _____

9.

1.2 m

34.5 cm

P = _____

A = _____

Solve each problem.

10. How many feet of fencing must the Smiths use to enclose a rectangular space that is 30 yards long and 28 feet wide?

11. Joe wanted to cover a rectangular trivet with 1-inch by 1-inch tiles. How many tiles would he need to cover a 6-inch by 9-inch trivet?

12. What is the perimeter of a rectangular flower bed 21 feet wide and 28 feet long?

13. A can of paint will cover 200 square feet. If each wall is 15 feet by 10 feet, how many cans of paint will Mai need to paint 3 walls?

14. The distance around a circle is called its **circumference** and the distance across a circle is called its **diameter**. For every circle, the ratio of the circumference to the diameter has the same value, approximately 3.14. That value is represented by the Greek letter π (pi). If r is the radius of a circle, you can use a formula to find the circumference C or the area A:

$$C = 2 \times \pi \times r \qquad\qquad A = \pi \times r \times r = \pi r^2$$

Find the circumference and the area of a circle with a radius of 15 centimeters.

Application: Home Remodeling

Jeff and Marisa are remodeling their home. Here are some of the problems and questions they have encountered.

1. On the roof, Jeff wants to put shingles on a rectangle 25 feet by 20 feet. What is the area of that part of the roof? What is the perimeter of that part of the roof?

2. Jeff bought enough shingles to cover 575 square feet. How many square feet of shingles will be left over after he covers the 25-foot by 20-foot rectangle?

3. Marisa is installing new windows. Each window is 2 feet by 4 feet. She installed 120 square feet of windows. How many windows did she install?

4. Jeff and Marisa put siding on their house. The siding they bought comes in 12-foot lengths. The front of their house is 32 feet wide. If they use three strips of siding to go across the front of the house, how much extra is there?

5. The strips of siding are 12 feet by 6 inches. How many strips will they use to cover 600 square feet?

6. Marisa bought scraps of plywood to build a dog house. The pieces were 2 feet wide, and their lengths were $3\frac{1}{2}$ feet, $2\frac{1}{4}$ feet, and 6 feet. What was the total length of the scraps of plywood? (Your answer should be in feet and inches.)

7. Asphalt for covering the driveway comes in 5-gallon buckets. Each bucket can cover 100 square yards. Their driveway is 100 feet long and 18 feet wide. How many buckets will they need? (Hint: 9 sq ft = 1 sq yd)

8. The angle formed by the two parts of the roof is 150°. Draw an angle of 150°. Is that angle acute, right, or obtuse?

9. In the attic, the chimney forms an angle of 75° with the floor. Draw an angle of 75°. Then find the complement and the supplement of the angle.

10. On the stairs, the handrail forms an angle of 50° with the vertical posts. Draw an angle of 50°. Then draw and label the complement and the supplement of a 50° angle.

Here are "before" and "after" plans for Jeff and Marisa's remodeling project.

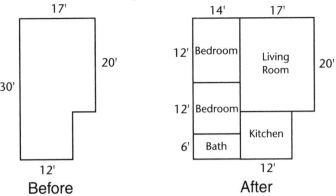

11. What is the perimeter of the house before the remodeling?

_____

12. What is the perimeter of the house after the remodeling?

_____

13. How many square feet were added to the house?

_____

14. Looking at the "after" picture, find the area of the living room, bathroom, and kitchen.

_____

15. Looking at the "after" picture, write the measurements for the living room and kitchen in yards.

Living Room _____

Kitchen _____

16. Looking at the "before" and "after" pictures of the remodeled house, what rooms were made in the new part of the house?

_____

Application: Scale on Drawings and Maps

On a scale drawing of an object, lengths in the picture or drawing represent actual lengths of the object. For example, if the scale for a drawing of a building is "1 inch = 4 feet," then 1 inch on the drawing represents 4 feet in the actual building.

A scale on a drawing or a map lets you calculate actual distances by using measurements from the drawing or map.

Example 1

Jana was looking at a map with a scale of "1 inch = 150 miles." On the map, the distance between Chicago and her hometown is 3 inches. What is the distance between her hometown and Chicago?

Step 1. Write the scale as a ratio.

$$\frac{1 \text{ inch}}{150 \text{ miles}}$$

Step 2. Set up a proportion using two ratios. Use "?" to represent the unknown distance.

$$\frac{1}{150} = \frac{3}{?}$$

Step 3. Cross multiply and solve for the unknown distance.

$$1 \times ? = 3 \times 150$$
$$? = 450$$

Answer: Jana's hometown is about 450 miles from Chicago.

Use each scale and measurement to find the actual length.

1. Scale: 1 in. = 6 ft
   Measurement: 6 in.
   Actual size: _____

2. Scale: 3 in. = 25 miles
   Measurement: 12 in.
   Actual size: _____

3. Scale: 2 in. = 75 ft
   Measurement: 10 in.
   Actual size: _____

4. Scale: 2 cm = 70 m
   Measurement: 15 cm
   Actual size: _____

5. Scale: 1 in. = 30 ft
   Measurement: $4\frac{1}{2}$ in.
   Actual size: _____

6. Scale: 1 cm = 10 km
   Measurement: 12 cm
   Actual size: _____

Use the scale "1 inch = 30 miles" for each problem.

7. On the map, the distance between Omaha and Des Moines is $4\frac{1}{2}$ inches. How far apart are Omaha and Des Moines?

8. The distance on the map between Kansas City and Omaha is 6 inches. How many miles is that?

Section 2 Cumulative Review

Fill in each blank with the correct measurement.

1. 1 yd = _____ in.

2. _____ ft = 1 mi

3. 1 cm = _____ mm

4. A straight angle contains _____°.

5. 1 ft = _____ in.

6. 1 m = _____ cm

7. A right angle contains _____°.

8. _____ m = 1 km

9. _____ ft = 1 yd

10. The complement of 30° is _____°.

Mark each measurement on the ruler.

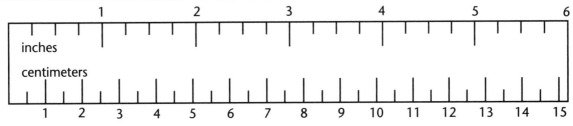

11. $3\frac{1}{4}$ in.

12. 2 cm 9 mm

13. $2\frac{15}{16}$ in.

14. 67 mm

15. 5 cm

16. 4.8 cm

Rewrite each measurement so it uses the given unit.

17. 18 in. = _____ yd

18. 6 m = _____ cm

19. 650 mm = _____ cm

20. 90 in. = _____ ft

21. $3\frac{1}{2}$ ft = _____ in.

22. $8\frac{1}{2}$ yd = _____ ft

23. 108 in. = _____ yd

24. $7\frac{1}{2}$ ft = _____ yd

25. 250 cm = _____ m

26. 3.25 m = _____ cm

Write the measurements from longest to shortest.

27. 38 in., 3 ft, 1 yd + 6 in., 2 yd _____

28. 18 cm, 18 m, 1600 cm, 80 m _____

Use a single unit to find each answer.

29. (6 ft + 8 in.) + (3 ft + 7 in.) = _____   31. $3\frac{1}{2}$ mi ÷ 7 = _____

30. (3 yd + 2 ft) − (1 yd + 17 in.) = _____   32. 5 × 7.2 mi = _____

Label each angle as acute, right, obtuse, or straight. Then use a protractor to measure the angle.

33.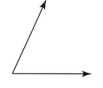

_____

35.

_____

34.

_____

36.

_____

Find the complement of each angle.

37. 62° _____   39. 80° _____   41. 1° _____

38. 17° _____   40. 15° _____   42. 45° _____

Find the supplement of each angle.

43. 115° _____   45. 120° _____   47. 175° _____

44. 40° _____   46. 10° _____   48. 15° _____

Use a protractor to draw the following angles.

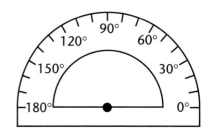

49. 30°   **52.** 45°

50. 150°   **53.** 135°

51. 90°   **54.** 150°

10 cm

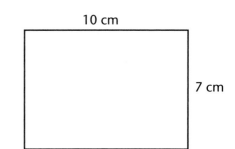

7 cm

Solve each problem.

55. Find the perimeter and
the area of the rectangle.

56. Find the circumference and area of a circular
garden with a 10-foot diameter.

57. The scale on a blueprint states "1 inch = 20 feet." On the blueprint, one room
is $7\frac{1}{2}$ inches long. What is the actual length of the room?

58. Jane needs $3\frac{1}{2}$ yards of fabric to make an outfit for the school play. How many
yards will she need to make eight outfits?

The Health Professions

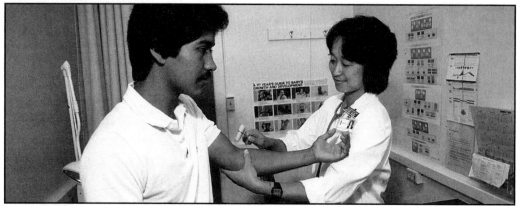

T he women and men who work in hospitals, trauma centers, doctors' offices, hospices, and other types of clinics are part of the health professions. All the nurses, technicians, doctors, and other workers help take care of people, and their jobs rely on many types of measurements.

Think about people being admitted into a hospital or medical clinic. The first medical person they see may be a paramedic or a nurse. These workers may measure temperature, weight, height, and blood pressure, and record the measurements in charts.

Doctors use these measurements and other measurements from their own examinations to prescribe courses of treatment. Pharmacists accurately measure dosages and count pills. Dietitians carefully measure amounts of food and prepare it at precise temperatures.

Think about many of the other workers in the health professions and how they use measurements: registered nurses who prepare IV (intravenous) solutions; laboratory technicians who draw blood and run tests; licensed practical nurses who may measure bandages and change dressings; and the workers who maintain the building's temperature and service the elevators.

All these workers need to take accurate measurements to help people get well and stay healthy.

Place yourself on a medical team. Start your diagnosis of a patient with the following questions. For each question, ask yourself what you would measure to answer the question. What type of measurement tool would you use?

- How much does the patient weigh?

- What is the patient's temperature?

- What is the patient's normal temperature?

- What is the patient's blood pressure?

- What medicine should the patient take? How large a dosage? How often?

- What foods should the patient be given from the different food groups? How many ounces or grams of each type of food?

> You can see that many of the measurements in the health professions involve weight and temperature. In these lessons, you will learn about different systems of units for measuring weight and for measuring temperature. Also, you will investigate several measuring tools for weight and temperature and apply those tools to solve problems.

Career Corner

Here are some of the jobs in the medical field that were mentioned.

admitting nurses	custodians	pharmacists
aides	licensed practical nurses	physicians
clerks	paramedics	registered nurses
cooks	dietitians	technicians

Can you think of any others?

Customary Units of Weight

What units do you use when you weigh something? When you use **ounces**, **pounds**, and **tons** to weigh an object, you are using customary units of weight. An ounce is a very small customary unit of weight, and a ton is a very large unit. Here are the relationships among these units.

$$1 \text{ pound (lb)} = 16 \text{ ounces (oz)}$$
$$1 \text{ ton (T)} = 2{,}000 \text{ lb}$$

Here are some examples of approximate weights.

1 oz

1 lb

1 T

Name three objects that might weigh the given amount.

1. 1 ounce (1 oz)

_____

2. 1 pound (1 lb)

_____

3. 1 ton (1 T)

_____

For each exercise, circle the heaviest object and underline the lightest object.

4. an orange, a melon, a raspberry

5. a garbage truck, a mini-van, a sports car

6. a glass of juice, a gallon of milk, a liter of juice

7. a safety pin, a button, a scissors

8. a 31-inch television, a portable radio, a CD player

Here are some common tools or scales used to measure weights. Some are used to measure very light objects, and others measure very heavy objects. Which of these scales have you used?

  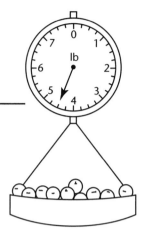

Example 1

This scale is in the produce section of a grocery store. Reading the scale is almost like reading a ruler. The scale is labeled in pounds. Each pound is divided into 4 parts, so each tick mark represents 4 ounces. Tai bought some tomatoes and used this scale to weigh them. How much do they weigh?

Step 1. Decide which units to use for the measurement.

__ lb and __ oz

Step 2. The arrow is between 4 and 5 pounds. Use the smaller number.

4 lb and __ oz

Step 3. Count the number of ounces more than 4 pounds.

4 lb and 8 oz

Answer: The tomatoes weigh 4 pounds 8 ounces.

Write the measurement shown on each scale.

9.

10.

11.

_____   _____   _____

Mark the given weight on each scale.

12.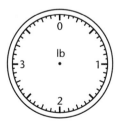

13.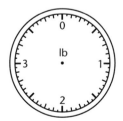

14.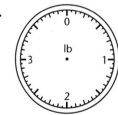

2 lb 15 oz   1 lb 0 oz   0 lb 7 oz

Example 2

A dietetic scale is used to measure small portions of food. The scale is labeled in ounces and half-ounces. How much does this portion of food weigh?

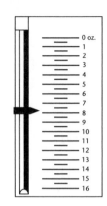

Step 1. Decide which unit to use for the measurement. ___ oz

Step 2. The pointer is between $7\frac{1}{2}$ and 8. Round the weight to the nearest ounce. 8 oz

Answer: The portion weighs 8 ounces.

Write the measurement shown on each scale.

15.
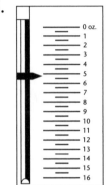

_____

16.
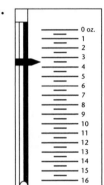

_____

17.
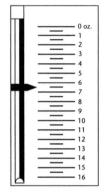

_____

Mark the given weight on each scale.

18.
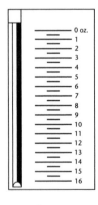

$2\frac{1}{2}$ oz

19.
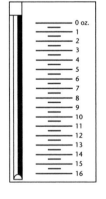

7 oz

20.
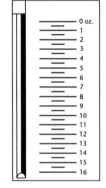

$4\frac{1}{2}$ oz

Metric Units of Weight

Have you ever noticed the weight measurements on a vegetable can? Often there are two measurements on the can, one in ounces and one in grams. A **gram** is a metric unit of weight. Two other metric units of weight are the **milligram** and the **kilogram**.

> 1,000 milligrams (mg) = 1 gram (g)
> 1,000 g = 1 kilogram (kg)

In the metric system of measurement, the prefix *milli-* always means $\frac{1}{1000}$ or one-thousandth. The prefix *kilo-* always means 1,000. A very light object is weighed in milligrams, and a heavy object is weighed in kilograms.

needle
about 300 mg

peanut
about 1 gram

telephone book
about 1 kg

Name three things that are sold, measured, or packaged in the given unit.

1. milligrams _____

2. grams _____

3. kilograms _____

Circle the better estimate for the weight of each object.

4. tennis ball   400 mg or 400 g

5. bag of apples   3 kg or 3 mg

6. straight pin   200 mg or 200 g

7. 4-year-old girl 15 g or 15 kg

8. sofa   40 mg or 40 kg

9. firecracker   30 kg or 30 g

10. motorcycle   220 g or 220 kg

11. stocking hat   800 kg or 800 g

12. feather   100 g or 100 mg

13. horse   400 kg or 400 g

Estimating in Ounces, Pounds, and Tons

This lesson shows four different tools for measuring weight. Some tools are used for light objects and some are used for heavy objects.

Scale 1

Scale 2

Scale 1 can be used for portions of food or other light objects. It uses ounces as a unit. Scale 2 often is used to weigh vegetables and fruit. This scale is labeled in pounds. Each pound is divided into 4 parts, so each tick mark represents 4 ounces.

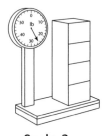

Scale 3

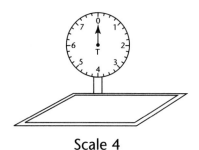

Scale 4

Scale 3 can be used to measure the weight of a package. The scale is labeled in increments of 10 pounds. Each 10-pound increment is divided into 5 parts, so each tick mark represents 2 pounds. Scale 4 is used to measure the weight of cars, trucks, and other heavy objects. The scale is labeled in tons. Each ton is divided into 4 parts, so each tick mark represents 500 pounds.

Example

Estimate the weight of a 31-inch TV. Then draw an appropriate scale for that weight. Show your estimated number on the scale.

Step 1. Think of an appropriate weight.

about 35 lb

Step 2. Draw and label an appropriate scale for measuring an object with that weight.

Answer: The scale shows 35 pounds.

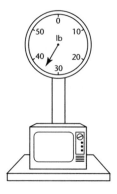

Estimate the weight of each object. Draw an appropriate scale for that weight. Then show your estimate on the scale.

1. a birthday card to be mailed overseas

2. a crate of canned goods

3. a horse

4. a bag of onions

5. a bag of oranges

6. a stove and a refrigerator

7. a portion of vegetables to be cooked

8. one serving of fish

9. an infant

10. a pickup truck

11. a small package to be mailed

12. a microwave

Changing Among Ounces, Pounds, and Tons

What happens to the weight of an object when you use a different unit to measure it? You know the weight of the object does not change. You will see that if you change from a smaller unit to a larger one, the number of units will decrease. If you change from a larger unit to a smaller one, the number of units will increase.

Here are the relationships among ounces, pounds, and tons.

> 1 pound (lb) = 16 ounces (oz)
> 1 ton (T) = 2,000 lb

Example 1

Jackie bought a 5-pound bag of rice. How many ounces of rice did she buy?

Step 1.	Write the problem.	5 lb = ____ oz
Step 2.	Write the relationship between ounces and pounds.	1 lb = 16 oz
Step 3.	We want to go from 1 pound to 5 pounds. So multiply both sides of the equation in Step 2 by 5.	5 lb = 80 oz
Answer:	Jackie has 80 ounces of rice.	

Rewrite each measurement so it uses the given unit.

1. 6 lb = ____ oz

2. 5 T = ____ lb

3. 2 lb = ____ oz

4. 2 T = ____ lb

5. 3 T = ____ lb

6. 10 lb = ____ oz

7. 20 T = ____ lb

8. 18 lb = ____ oz

9. 4 lb = ____ oz

When you change from a smaller unit to a larger one, some of the smaller units may be left over. Here is an example.

Example 2

Change 56 ounces to pounds.

Step 1.	Write the problem.	56 oz = ____ lb
Step 2.	Write the relationship between ounces and pounds.	16 oz = 1 lb
Step 3.	Divide 56 by 16.	$\frac{56}{16} = 3\frac{8}{16} = 3\frac{1}{2}$
Answer:	56 ounces is $3\frac{1}{2}$ pounds.	

Rewrite each measurement so it uses the given unit.

10. 2,500 lb = _____ T

19. 32 oz = _____ lb

28. $1\frac{1}{4}$ lb = _____ oz

11. 84 oz = _____ lb

20. 20,000 lb = _____ T

29. 9.2 T = _____ lb

12. 22 oz = _____ lb

21. 16 oz = _____ lb

30. $2\frac{1}{10}$ T = _____ lb

13. 6,800 lb = _____ T

22. 8,000 lb = _____ T

31. $4\frac{1}{2}$ T = _____ lb

14. 3,600 lb = _____ T

23. 80 oz = _____ lb

32. 2.3 lb = _____ oz

15. 12 oz = _____ lb

24. 112 oz = _____ lb

33. $3\frac{1}{4}$ T = _____ lb

16. 68 oz = _____ lb

25. 24,000 lb = _____ T

34. $6\frac{1}{2}$ lb = _____ oz

17. 8 oz = _____ lb

26. 320 oz = _____ lb

35. $2\frac{3}{4}$ lb = _____ oz

18. 1,000 lb = _____ T

27. 40,000 lb = _____ T

36. 5.5 lb = _____ oz

Use equivalent measurements for ounces, pounds, and tons to solve each problem.

37. Sam drives a delivery truck. He delivered 2.6 tons of produce last week. How many pounds did he deliver?

38. The package Juan wanted to mail home weighed $12\frac{1}{2}$ pounds. How many ounces did the package weigh?

39. Mulch is sold by the ton. Fran's lawn service needs 2,800 pounds to complete a job. How many tons does she need?

40. A can of soup weighs 22 ounces. How many pounds is that?

41. Tru bought a $3\frac{1}{2}$-pound bag of oranges. How many ounces did he buy?

42. Sue wanted to make hamburgers from a 6-pound package of ground meat. How many 4-ounce patties can she make?

43. Tran weighed the bag of carrots as 64 ounces. How many pounds is that?

44. The city needed 22 tons of rock to cover the landscaping of the city park. How many pounds is that?

45. Kim delivered a truckload of cattle. If the weight of the cattle was 36,000 pounds, how many tons did the cattle weigh?

46. Chu bought $1\frac{3}{4}$ pounds of dried fruit. How many ounces did he buy?

47. Tia cooked 120 ounces of hamburger. How many pounds did she cook?

48. To figure out the cost to ship a package, George has a chart with pounds and ounces. Kai wanted to ship a 90-ounce package. For how many pounds must George charge Kai?

Estimating in Milligrams, Grams, and Kilograms

Here are the approximate weights, in milligrams, grams, and kilograms, of some common objects.

	milligrams	grams	kilograms
a vitamin tablet	1,000 mg	1 g	0.001 kg
a dinner roll	30,000 mg	30 g	0.03 kg
a can of corn		500 g	0.5 kg
a box of laundry soap		1,000 g	1 kg
a large horse		1,000,000 g	1,000 kg

Complete each measurement by writing mg, g, or kg.

1. a hot dog   50 ____

2. a refrigerator   100 ____

3. a small steak   200 ____

4. a brick   1 ____

5. a compact disc   500 ____

6. a tennis ball   175 ____

7. a mini-van   900 ____

8. a cereal serving   30 ____

9. a loaf of bread   0.5 ____

10. a postage stamp   0.1 ____

11. a sandwich   0.25 ____

12. a melon   6 ____

13. a bag of apples   2 ____

14. a stick of gum   1 ____

15. a newspaper   0.4 ____

16. a truck of cement   3,000 ____

Circle the most appropriate unit for measuring the weight of each object.

17. a business envelope	mg	g	kg
18. a jar of salsa	mg	g	kg
19. sodium in a bowl of popcorn	mg	g	kg
20. an elephant	mg	g	kg
21. a glass of milk	mg	g	kg
22. a bag of potatoes	mg	g	kg
23. an empty shipping crate	mg	g	kg

Changing Among Milligrams, Grams, and Kilograms

You saw what happened when you used different customary units to weigh an object. The same is true using metric units: if you change from a smaller unit to a larger one, the number of units will decrease. If you change from a larger unit to a smaller one, the number of units will increase.

Here are the relationships among milligrams, grams, and kilograms.

$$
\begin{aligned}
1 \text{ mg} &= 0.001 \text{ g} \\
1{,}000 \text{ mg} &= 1 \text{ g} \\
1 \text{ g} &= 0.001 \text{ kg} \\
1{,}000 \text{ g} &= 1 \text{ kg}
\end{aligned}
$$

Example 1

A cleaning crew needs to use 2 kilograms of powdered disinfectant to prepare the surgery suite of a hospital. How many grams of disinfectant are they going to use?

Step 1.	Write the problem.	2 kg = ___ g
Step 2.	Write the relationship between kilograms and grams.	1 kg = 1,000 g
Step 3.	We want to go from 1 kilogram to 2 kilograms. So multiply both sides of the equation in Step 2 by 2.	2 kg = 2,000 g

Answer: They will use 2,000 grams of disinfectant.

Rewrite each measurement so it uses the given unit.

1. 6 kg = ___ g

2. 27 kg = ___ g

3. 18 g = ___ mg

4. 17 kg = ___ g

5. 245 g = ___ mg

6. 3 g = ___ mg

Sometimes a measurement contains a mixed number. Here is an example.

Example 2

Change $2\frac{1}{2}$ kilograms to grams.

Step 1.	Write the problem.	$2\frac{1}{2}$ kg = ___ g
Step 2.	Write the relationship between kg and g.	1 kg = 1,000 g
Step 3.	We want to go from 1 kilogram to $2\frac{1}{2}$ kilograms. So multiply each side of the equation in Step 2 by $2\frac{1}{2}$.	$2\frac{1}{2}$ kg = 2,500 g
Answer:	$2\frac{1}{2}$ kg = 2,500 g	

Rewrite each measurement so it uses the given unit.

7. 3.6 g = _____ mg

8. 8.6 kg = _____ g

9. $6\frac{3}{10}$ kg = _____ g

10. $4\frac{1}{4}$ kg = _____ g

11. 12.6 g = _____ mg

12. 2.45 kg = _____ g

13. $6\frac{2}{5}$ g = _____ mg

14. 7.25 g = _____ mg

— Example 3 —

Ethel wants to reduce the amount of sodium in her diet. Two whole pizzas have 2,500 milligrams of sodium. How many grams of sodium do the pizzas contain?

Step 1.	Write the problem.	2,500 mg = _____ g
Step 2.	Write the relationship between g and mg.	1,000 mg = 1 g
Step 3.	We want to go from 1,000 milligrams to 2,500 milligrams. So multiply both sides of the equation in Step 2 by 2.5.	2,500 mg = 2.5 g

Answer: The pizzas contain 2.5 grams of sodium.

Rewrite each measurement so it uses the given unit.

15. 20,000 g = _____ kg

16. 6,000 mg = _____ g

17. 7,000 mg = _____ g

18. 8,000 g = _____ kg

19. 2,000 g = _____ kg

20. 26,000 mg = _____ g

21. 2,100 mg = _____ g

22. 4,725 mg = _____ g

23. 2,750 g = _____ kg

24. 4,200 g = _____ kg

25. 325 mg = _____ g

26. 3,600 g = _____ kg

Use the equivalent measurements for milligrams, grams, and kilograms to solve each problem.

27. The frozen dinner has $3\frac{1}{2}$ grams of fat. How many milligrams of fat does it contain?

28. George bought a 2.8-kilogram package of ground beef. How many grams did he buy?

29. Sheila weighed a portion of vegetables. It weighed 90 grams. How many milligrams is the portion of vegetables?

30. Dietary guidelines say a person on a 2,500-calorie-per-day diet should consume fewer than 24 grams of sodium daily. How many milligrams is that?

31. How many 500-gram packages of rice can you make from a 4-kilogram box of rice?

32. Kim bought 1.75 kilograms of candy. How many grams is that?

33. Mai read the label on a box of cereal. It said the daily recommended amount of potassium is 3,500 milligrams. How many grams is that?

34. Hector weighed a piece of chicken at 100 grams. How many kilograms is that?

35. During the week, James figured he had eaten about 35,000 milligrams of fat. How many grams is that?

36. Joi is a shipping clerk. She marked a package as 1,750 grams. How many kilograms is that?

37. A packing company put glasses in very small boxes. Each box weighed 3,500 grams. How many kilograms did each box weigh?

38. Sara plans to eat at least 32,500 milligrams of fiber daily. How many grams of fiber does she plan to eat daily?

Adding and Subtracting Weights

When you work with measurements, sometimes you need to add or subtract the measurements. Adding or subtracting measurements can involve regrouping and borrowing.

── Example 1 ──────────────────────

The Tran's baby girl has gained 2 pounds 7 ounces since her birth. If she weighed 6 pounds 12 ounces when she was born, how much does she weigh now?

Step 1.	Line up the measurements, putting like units under like units.	6 lb 12 oz + 2 lb 7 oz
Step 2.	Add the ounces and add the pounds.	6 lb 12 oz + 2 lb 7 oz 8 lb 19 oz
Step 3.	Change 19 ounces to pounds and ounces.	19 oz = 16 oz + 3 oz = 1 lb 3 oz
Step 4.	Rewrite the total.	8 lb + 19 oz = 8 lb + 1 lb + 3 oz = 9 lb 3 oz

Answer: The baby now weighs 9 pounds 3 ounces.

── Example 2 ──────────────────────

Subtract 9 pounds 17 ounces from 13 pounds 4 ounces.

Step 1.	Line up the measurements, putting like units under like units.	13 lb 4 oz − 9 lb 17 oz
Step 2.	Think of 13 lb as 12 lb 16 oz. Then rewrite 13 lb 4 oz as 12 lb 20 oz.	12 20 ~~13~~ lb ~~4~~ oz − 9 lb 17 oz
Step 3.	Subtract the pounds and subtract the ounces.	12 lb 20 oz − 9 lb 17 oz 3 lb 3 oz

Answer: The difference is 3 pounds 3 ounces.

Add or subtract.

1. 6 lb 13 oz + 4 lb 7 oz

2. 4 lb 2 oz – 2 lb 12 oz

3. 18 lb 12 oz + 6 lb 9 oz

4. 7 lb 12 oz – 3 lb 4 oz

5. 12 lb 7 oz – 8 lb 2 oz

6. 2 lb 7 oz + 4 lb 3 oz

7. 3 lb 6 oz + 7 lb 15 oz

8. 2 lb 4 oz – 1 lb 10 oz

9. 16 lb 2 oz – 10 lb 4 oz

10. 12 lb 2 oz + 4 lb 3 oz

11. 8 lb – 3 lb 7 oz

12. 2 lb 14 oz + 1 lb 7 oz

To add or subtract metric measurements, make sure the measurements use the same units.

Example 3

Find the sum and the difference of 5.32 kilograms and 150 grams.

Step 1.	Rewrite the measurements using the same unit.	5.32 kg = 5,320 g	
		Sum	**Difference**
Step 2.	Line up the two	5,320	5,320
	measurements. Then	+ 150	– 150
	add or subtract.	5,470	5,170

Answer: The sum is 5,470 grams (or 5.47 kilograms).
The difference is 5,170 grams (or 5.17 kilograms).

Find each sum or difference.

13. 36 g + 2 kg

14. 15 g – 400 mg

15. 12 kg – 750 g

16. 12 mg + 12 g

17. 1 kg + 1 g + 1 mg

18. 1 kg + 1,000 g + 1,000,000 mg

Solve each problem.

19. If James removes 6 pounds 4 ounces from a container that weighs 18 pounds 12 ounces, how much is left?

20. Carlos bought two packages of hamburger. One weighed 2 pounds 7 ounces and the other weighed 1 pound 12 ounces. How much hamburger did he buy?

Multiplying and Dividing Weights

Sometimes you need to multiply or divide weights. If a large carton contains many of the same item, you can use division to find the weight of each item. If you are shipping many packages that have the same weight, you can use multiplication to find the total weight.

Example 1

Rachel bought a "family pack" of lean ground beef that contains 5 packages. Each package weighs 4 pounds 9 ounces. What is the total weight of the ground beef?

Step 1. Decide which math operation to use.

The operation to use is multiplication.

Step 2. Write the problem.

$$\begin{array}{r} 4 \text{ lb } 9 \text{ oz} \\ \times \quad 5 \\ \hline \end{array}$$

Step 3. Multiply the ounces and the pounds by 5.

$$\begin{array}{r} 4 \text{ lb } 9 \text{ oz} \\ \times \quad 5 \\ \hline 20 \text{ lb } 45 \text{ oz} \end{array}$$

Step 4. 45 ounces is the same as 32 ounces plus 13 ounces. So rewrite 20 pounds 45 ounces.

20 lb + 45 oz
= 20 lb + 2 lb + 13 oz
= 22 lb + 13 oz

Answer: The total weight is 22 pounds 13 ounces.

Example 2

Rachel's family has a favorite recipe that uses 3.5 pounds of ground beef. How can Rachel use her calculator to find out how many times she can make that dish before she uses all 22 pounds 13 ounces of ground beef?

Step 1. Decide which math operation to use.

The operation to use is division.

Step 2. Use a calculator to write 22 pounds 13 ounces as a decimal.

22 lb 13 oz = $22 + \frac{13}{16}$
= 22 + 0.8125
= 22.8125

Step 3. Use a calculator to divide 22.8125 by 3.5.

22.8125 ÷ 3.5 = 6.5179

Answer: Rachel has enough ground beef to make the recipe 6 times. (She will have enough left over for another half of the recipe.)

Multiply or divide.

1. 2 lb 5 oz × 3

2. 4 lb 5 oz ÷ 3

3. 1 lb 3 oz × 8

4. 6 lb 2 oz ÷ 2

5. 12 lb 3 oz ÷ 5

6. 7 lb 10 oz × 4

7. 6 oz × 4

8. 3 lb 8 oz ÷ 2

9. 12 lb 5 oz × 6

10. 22 lb 2 oz ÷ 6

11. 24 lb 7 oz × 4

12. 38 lb 1 oz ÷ 7

13. 10 lb 1 oz × 3

14. 6 lb 12 oz ÷ 6

Solve each problem.

15. A box of detergent weighs 40 pounds 10 ounces. Joe wants to put it into 5 equal packages. How much should each package weigh?

16. Elena is mailing the same present to 7 people. If the present weighs 2 pounds 3 ounces, how much will all 7 packages weigh?

17. Henri bought 7 pounds 12 ounces of ham to feed 31 people. How many ounces can he serve each person?

18. Clarence bought 5 cans of chili beans. Each can weighed 1 pound 2 ounces. What was the total weight of chili beans?

Comparing and Ordering Weights

Sometimes you want to know which of two objects is lighter or heavier. It is easier to compare weights if you change the measurements to the same unit. When you compare measurements, you can use the math symbols < or > to represent "is less than" or "is greater than."

Example 1

Maria has a 2-pound package of rice. She wants to use a recipe that calls for 24 ounces of rice. Does she have enough rice to make the recipe?

Step 1.	Write the two measurements.	2 lb; 24 oz
Step 2.	Write the relationship between pounds and ounces.	1 lb = 16 oz
Step 3.	Change one of the measurements.	2 lb = (2)(16) = 32 oz
Step 4.	Compare the measurements.	32 oz > 24 oz

Answer: Maria has 32 ounces of rice, which is enough rice for the recipe.

Example 2

Which is smaller, 3,500 milligrams or 3 grams?

Step 1.	Write the two measurements.	3,500 mg; 3 g
Step 2.	Write the relationship between milligrams and grams.	1,000 mg = 1 g
Step 3.	Change one of the measurements.	$3,500 \text{ mg} = \dfrac{3,500}{1,000} = 3.5 \text{ g}$
Step 4.	Compare the measurements.	3 g < 3.5 g

Answer: 3 grams is smaller than 3,500 milligrams.

Circle the smaller measurement.

1. 3 lb or 50 oz

2. 3.5 kg or 4,200 g

3. $2\frac{1}{2}$ lb or 42 oz

4. 3 T or 4,800 lb

5. 128 oz or $7\frac{1}{2}$ lb

6. 2,700 mg or 2.5 g

7. 18 oz or 2 lb

8. 4 kg or 360 g

Fill in each blank with >, <, or =.

9. 6 oz ____ 1 lb

10. 2 T ____ 800 lb

11. 420 mg ____ 3 g

12. 3 kg ____ 3,000 g

13. $2\frac{1}{2}$ T ____ 7,000 lb

14. 124 oz ____ 7 lb

15. 3 kg ____ 300 mg

16. 6 lb ____ 96 oz

Write the measurements from lightest to heaviest.

17. 14 oz, 1.5 lb, 7 oz, 2 lb

_____

18. 18 lb, 3 T, 270 lb, 1.6 T

_____

19. 16 mg, 6 kg, 6 g, 482 g

_____

20. 800 g, 240 mg, 1 kg, 45 g

_____

Solve each problem.

21. Kayla bought a 46-ounce bag of potatoes. Is that greater than or less than 3 pounds?

22. The doctor said the newborn weighed $7\frac{1}{2}$ pounds. The nurse said the newborn weighed 7 pounds 8 ounces. Are both weights the same?

23. Sam, a shipping clerk, is weighing two small packages for mailing. Package A weighs 22 ounces; package B weighs 1 pound 4 ounces. Which package weighs more?

24. A nutrition label on a can of mixed nuts says the fat content is 9 grams per serving and the sodium is 400 milligrams per serving. Which amount is smaller?

Comparing Customary and Metric Units of Weight

Sometimes the weight of an object is given in both customary units and in metric units. For example, a candy bar wrapper might list the weight as

net wt. 1.6 oz (45 g).

Ounces, pounds, and tons are the basic units of weight in the customary measurement system. Milligrams, grams, and kilograms are the basic units of weight in the metric system.

Here are some equivalent measurements for customary and metric units of weight.

16 oz = 1 lb	1,000 mg = 1 g	1 kg ≈ 2.2 lb	
2,000 lb = 1 T	1 mg = 0.001 g	1 oz ≈ 28 g	
	1,000 g = 1 kg	1 T ≈ 909 kg	

Example 1

Jim's son wanted to know which is more, an ounce or a gram. How could Jim answer the question?

Step 1.	Write the two units.	1 oz 1 g
Step 2.	Write the relationship between the two units.	1 oz ≈ 28 g
Step 3.	Decide which is larger.	It takes 28 grams to make 1 ounce, so 1 gram is much smaller than 1 ounce.

Answer: An ounce is larger than a gram.
(An ounce is about 28 times as large as a gram.)

Circle the larger measurement.

1. 1 gram   or   1 lb

2. 1 ton   or   1 lb

3. 1 kg   or   1 lb

4. 1 g   or   1 oz

5. 1 oz   or   1 lb

6. 1 ton   or   1 g

7. 1 oz   or   1 kg

8. 1 mg   or   1 g

9. 1 kg   or   1 ton

10. 1 mg   or   1 oz

Name three objects whose weight falls between the given amounts.

11. 1 gram and 50 grams

_____

12. 1 ton and 5 tons

_____

13. 1 pound and 5 pounds

_____

14. 1 kilogram and 5 kilograms

_____

15. Where might you see weight of an object listed using both customary units and metric units?

_____

Circle the more appropriate measurement for the weight of each object.

16. a strawberry
    6 oz or 6 g

17. a bowling ball
    3 oz or 3 kg

18. a can of green beans
    1 g or 1 lb

19. a child
    24 kg or 24 oz

20. a cow
    1 kg or 1 ton

21. a paper clip
    400 mg or 400 g

Use the information on customary and metric units of weight to answer each question.

22. An object weighs 14 grams. What fraction of an ounce does it weigh?

23. A book weighs 1 pound. Does it weigh more or less than half a kilogram?

24. Is 1,000 ounces more or less than 1 kilogram?

25. Is 200 grams more or less than 1 pound?

26. About how many pounds does it take to balance
    1 kilogram + 1 gram + 1 milligram?

27. About how many kilograms does it take to balance
    1 ton + 1 pound + 1 ounce?

Focus on Thermometers: Fahrenheit and Celsius Scales

You probably have seen the two different thermometers that are used to measure temperature. One measures the temperature in degrees Fahrenheit (°F). The other measures the temperature in degrees Celsius (°C).

This thermometer shows both the Fahrenheit and Celsius scales. Some of the most common measurements are shown.

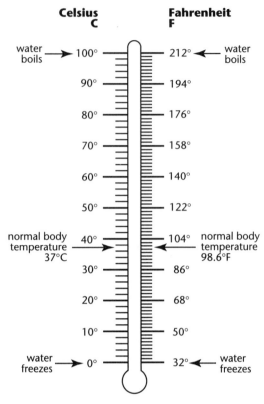

— Example 1 —

At what temperature does water freeze?

Step 1. Find the two labels in the diagram that say "water freezes."

Step 2. Look at the labels on the top of each scale for the words "Celsius" and "Fahrenheit."

Answer: When water freezes, the reading on a Celsius thermometer shows 0°C. The reading on a Fahrenheit thermometer shows 32°F.

Use the Celsius and Fahrenheit scales on the thermometer to answer each question.

1. What is normal body temperature?

   _____ °C   _____ °F

2. At what temperature does water boil?

   _____ °C   _____ °F

3. How many degrees are there between the freezing and boiling points of water?

   _____ °C   _____ °F

Use the thermometer above to fill in the chart.

	°F	°C
4.		30°C
5.		80°C
6.	104°F	
7.	194°F	
8.		50°C

Circle the warmer temperature.

9. 28°C or 28°F

11. 72°F or 24°C

10. 90°C or 90°F

12. 60°F or 48°C

Circle the more reasonable temperature for each object.

13. ice cream
    30°C or 30°F

17. sleet
    28°F or 28°C

14. swimming pool water
    70°C or 70°F

18. a cool autumn day
    8°F or 8°C

15. comfortable
    sleeping room
    68°C or 68°F

19. normal body
    temperature
    98.6°F or 98.6°C

16. a cup of hot tea
    80°C or 80°F

20. a cold winter's day
    12°C or 12°F

Circle the colder temperature.

21. 6°C or 6°F

23. 29°C or 29°F

25. 75°C or 104°F

22. 0°C or 0°F

24. 80°C or 150°F

26. 20°C or 80°F

In the diagram below, the temperature 14°C is marked on the thermometer. Mark the given temperatures on the thermometer.

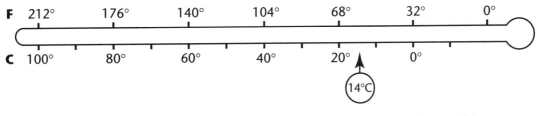

27. 70°F

28. 180°F

29. 65°C

30. 42°C

Focus on Thermometers: Temperatures

Sometimes the radio or television reports the current temperature and tells you there will be a particular temperature change. You can use addition or subtraction to find the expected, changed temperature.

Example 1

A weather report, showing this thermometer, says that the morning temperature is 40°F and that the temperature will fall 8 degrees during the day. What is the temperature expected at the end of the day?

Step 1. Write the morning temperature.   40°F
Step 2. "Fall" means "become lower."   40 – 8
So "fall 8 degrees" means to subtract 8 from 40.
Step 3. Subtract.   40 – 8 = 32
Answer: The temperature at the end of the day is expected to be 32°F.

Find the result for each temperature change.

1. Current temperature: 72°F
   Change: 12° drop

2. Current temperature: 38°C
   Change: 10° increase

3. Current temperature: 36°F
   Change: 17° drop

4. Current temperature: 25°C
   Change: 6° drop

5. Current temperature: 92°F
   Change: 6° decrease

6. Current temperature: 14°C
   Change: 3° rise

On a cold day or night, the temperature may be below zero.

Example 2

A weather report, showing this thermometer, says that the evening temperature is 12°F and that the temperature will fall 17 degrees during the night. What is the coldest temperature expected during the night?

Step 1. Write the evening temperature.   12°F
Step 2. "Fall" means "become lower."   12 – 17
So "fall 17 degrees" means to subtract 17 from 12.
Step 3. Subtract.   12 – 17 = –5
Answer: The coldest temperature expected during the night is –5°F.

Find the result for each temperature change.

7. Current temperature: 5°C
   Change: 8° drop

8. Current temperature: –14°C
   Change: 7° increase

9. Current temperature: 0°F
   Change: 6° drop

10. Current temperature: 21°C
    Change: 6° drop

11. Current temperature: 0°F
    Change: 3° increase

12. Current temperature: 21°F
    Change: 36° drop

13. Current temperature: 18°F
    Change: 21° fall

14. Current temperature: 3°C
    Change: 7° drop

Circle the colder temperature. You can look at a thermometer to help you decide.

15. –8°F or 8°F

16. 15°C or –16°C

17. 100°F or 32°C
    (Be careful comparing
    °F and °C.)

18. 0°F or 6°F

19. 4°C or –7°C

20. 0°C or 30°F
    (Be careful comparing
    °F and °C.)

Mark the given temperatures on the Fahrenheit thermometer.

21. 36°F

22. –12°F

23. 48°F

24. 0°F

25. 45°F

26. –5°F

27. –17°F

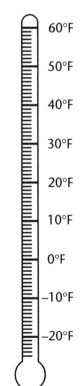

Focus on Thermometers: Changing Between °F and °C

There are two formulas for changing between Fahrenheit and Celsius temperatures.

Change from C to F:	Change from F to C:
$F = \frac{9}{5}C + 32$	$C = \frac{5}{9}(F - 32)$

___ Example 1 ___

The outdoor thermometer at the bank says the temperature is 25°C. Manuel wants to know what the Fahrenheit temperature is.

Step 1. Write the formula for changing Celsius to Fahrenheit. $\qquad F = \frac{9}{5}C + 32$

Step 2. Substitute the given value of C into the formula. $\qquad F = \frac{9}{5} \cdot 25 + 32$

Step 3. Find the value of F. $\qquad = \frac{9}{\overset{}{\underset{1}{5}}} \cdot \overset{5}{\cancel{25}} + 32$

$$= 45 + 32$$
$$= 77$$

Answer: 25°C = 77°F

Change each Celsius temperature to a Fahrenheit temperature.

1. 30°C = ____ °F   3. 75°C = ____ °F   5. 45°C = ____ °F

2. 40°C = ____ °F   4. 10°C = ____ °F   6. 90°C = ____ °F

___ Example 2 ___

The temperature is 50°F. What is the temperature on a Celsius thermometer?

Step 1. Write the formula for changing Fahrenheit to Celsius. $\qquad C = \frac{5}{9}(F - 32)$

Step 2. Substitute the given value of F into the formula. $\qquad C = \frac{5}{9}(50 - 32)$

Step 3. Find the value of C. $\qquad C = \frac{5}{9} \cdot 18$

$$= \frac{5}{\overset{}{\underset{1}{9}}} \cdot \overset{2}{\cancel{18}}$$
$$= 10$$

Answer: 50°F = 10°C

Change each Fahrenheit temperature to a Celsius temperature.

7. 104°F = _____ °C

8. 158°F = _____ °C

9. 68°F = _____ °C

10. 32°F = _____ °C

11. 72°F = _____ °C

12. 86° F = _____ °C

13. −4°F = _____ °C

14. 59°F = _____ °C

15. 212°F = _____ °C

16. −40°F = _____ °C

Use $F = \frac{9}{5}\,C + 32$ or $C = \frac{5}{9}\,(F - 32)$ to solve each problem.

17. Water boils at 100 degrees Celsius. What Fahrenheit temperature is that?

18. Water freezes at 0 degrees Celsius. What Fahrenheit temperature is that?

19. By putting salt on a sidewalk, snow will melt at or above 23 degrees Fahrenheit. What Celsius temperature is that?

20. An air conditioner is set at 77 degrees Fahrenheit. What Celsius temperature is that?

21. A heater's thermostat is set at 68 degrees Fahrenheit. What Celsius temperature is that?

22. The high temperature for the month of February was 59 degrees Fahrenheit. What Celsius temperature is that?

23. The storage refrigerator for beverages must be kept below 50 degrees Fahrenheit. What Celsius temperature is that?

24. A case of chocolate from Europe has a warning on the side: "Do not store above 30 degrees Celsius." At what Fahrenheit temperature should the chocolate be stored?

Application: Windchill

When the wind is blowing, a cold day feels even colder! This effect is called windchill. The following chart gives you the windchill temperature, or what the temperature feels like, for different wind speeds and different actual temperatures.

Windchill Table						
Actual Temperature (°F)	50°	40°	30°	20°	10°	0°
Wind speed (mph)						
5	48°	37°	27°	16°	7°	–5°
10	40°	28°	16°	3°	–9°	–22°
15	36°	22°	9°	–5°	–18°	–31°
20	32°	18°	4°	0°	–24°	–39°
25	30°	16°	1°	–15°	–29°	–44°
30	28°	13°	–2°	–18°	–33°	–49°
35	27°	11°	–4°	–20°	–35°	–52°
40	26°	10°	–5°	–21°	–37°	–53°

— Example 1 —

What is the windchill when the temperature is 10° and the wind speed is 15 mph?

Step 1. Locate "15 mph" along the left and "10°" along the top.

Step 2. Find the table entry where the "15 mph" row meets the "10°" column.

Answer: The windchill is –18°F. When the actual temperature is 10° and the wind is blowing at 15 miles per hour, the temperature feels like it is 18 degrees below zero on a calm day.

Use the windchill table to answer each question.

1. Are the temperatures given in °C or °F?

2. What is the lowest actual temperature listed on the chart?

3. If the temperature is 30°F and the wind speed is 15 miles per hour, what is the windchill?

Example 2

What combinations of actual temperature and wind speed result in a windchill of 16°?

Step 1. Locate "16°" in the table.
(Can you find all three entries?)

Step 2. From each entry of "16°," read the actual temperature and wind speed for that entry.

Answer: According to the windchill table, there are three conditions that give you a windchill of 16°:
actual temperature of 20°F, wind speed of 5 mph
actual temperature of 30°F, wind speed of 10 mph
actual temperature of 40°F, wind speed of 25 mph.

Use the windchill table to answer each question.

4. If the windchill is –2°F and the actual temperature is 30°F, what is the wind speed?

5. If the windchill is –37°F and the wind speed is 40 mph, what is the actual temperature?

6. At what wind speeds is the windchill colder than –50°F?

7. If the temperature is 20°F and the wind speed is 20 miles per hour, what is the windchill?

8. If the temperature is 10°F and the wind speed is 30 miles per hour, what is the windchill?

9. If the temperature is 0°F and the wind speed is 30 miles per hour, what is the windchill?

10. At what conditions is the windchill 18°F?

11. At what conditions is the windchill –24°F?

Application: Doctors' Offices

1. Lucy weighed 124 pounds on her last doctor's visit. The nurse weighed Lucy today and said she had gained 12 pounds. How much does Lucy weigh? Draw a weight scale and show that weight on the scale.

2. When the doctor saw Lucy's weight gain, he said she needed to lose 8 pounds. How much will she weigh after she loses the weight?

3. Jerry, a veterinarian, wanted to weigh his dog Tugger. Jerry weighed himself. The scale read 196 pounds. Jerry held Tugger and weighed himself again. The scale read 225 pounds. How much does Tugger weigh?

4. While Karma was pregnant, she gained weight every month. The nurse recorded her weight gain each month.

Jan.	2 lb 5 oz	Apr.	3 lb 1 oz	July	6 lb 3 oz
Feb.	2 lb 4 oz	May	3 lb 7 oz	Aug.	4 lb 10 oz
Mar.	1 lb 12 oz	June	2 lb 5 oz	Sept.	5 lb 15 oz

   a. What was her total weight gain? Record your answer in pounds and ounces.

   b. What was Karma's average weight gain over the 9-month period?

5. Karma had twin girls. When they were born, their total weight was 16 lb 6 oz. Their weight has tripled since their birth. How much do they now weigh?

6. What is the average weight of each twin now?

7. What was the average weight of each twin when they were born? Draw a weight scale and show that weight on the scale.

8. Jose was not feeling well. The nurse took his temperature and found out he had a fever of 102.5°F. Draw a thermometer and show that temperature on the thermometer. How much above 98.6°F was his temperature?

9. During an operation, doctors lowered Mai's body temperature. At the beginning of the operation, her body temperature was 94.9°F. Draw a thermometer and show that temperature on the thermometer. How much below 98.6°F was her temperature?

10. Tyler's temperature was 99.2°F at 10 A.M. and 101.9°F at 3 P.M. Draw a thermometer and show those temperatures on the thermometer. By how much had his temperature risen?

11. The Jones' baby was sick. When the nurse took his temperature, it was 104.2°F. Draw a thermometer and show that temperature on the thermometer. How much above 98.6°F was the baby's temperature?

12. Joyce's temperature was 102.4°F, so she took some medicine. Later it was 99.9°F. Draw a thermometer and show those temperatures on the thermometer. By how much did her temperature drop?

13. During a routine physical, Fred's temperature was normal at 97.4°F. Draw a thermometer and show that temperature on the thermometer. By how much does Fred's normal temperature differ from 98.6°F?

Section 3 Cumulative Review

Rewrite each measurement so it uses the given unit.

1. 1 lb = _____ oz

2. 1 kg ≈ _____ lb

3. 1 g = _____ mg

4. 1 T = _____ lb

5. 1 oz ≈ _____ g

6. 66 oz = _____ lb

7. 4,500 lb = _____ T

8. 3,500 mg = _____ g

9. $2\frac{1}{2}$ lb ≈ _____ g

10. 4.3 kg = _____ g

Circle the larger measurement.

11. 2 lb or 2 kg

12. 32°F or 32°C

13. 600 g or 600 mg

14. 0°C or 10°F

15. $2\frac{1}{2}$ tons or $2\frac{1}{2}$ kilograms

16. 15 oz or 15 g

Circle the better estimate for each object.

17. temperature of bath water   80°C or 80°F

18. weight of a can of soup   16 oz or 16 g

19. temperature of a fudge bar   26°C or 26°F

20. weight of a phone book   1 g or 1 kg

21. weight of an elephant   2.2 kg or 2.2 tons

Mark the given weight on each scale.

22.
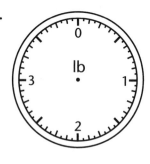

2 lb 3 oz

23.
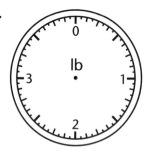

$1\frac{1}{2}$ lb

24.
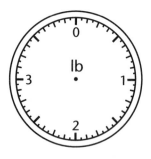

3 lb 13 oz

84   *WEIGHT AND TEMPERATURE*

Write the measurements from lightest to heaviest.

25. 26 g, 26 lb, 26 oz, 26 kg _____

26. 2 T, 2 oz, 2 lb _____

27. 1 kg, 1 mg, 1 g _____

Solve each problem.

28. Jason has to divide a 27-pound box of candy into 6 equal packages. What is the weight of each?

29. Jackie bought 2 pounds 7 ounces of ham and 3 pounds 4 ounces of turkey at the grocery store. How much meat did she buy?

30. The total weight of three packages is 4 pounds 5 ounces. If one package weighs 2 pounds 7 ounces, what is the total of the other two?

31. Chu bought 7 packages of cheese. Each package weighed 12 ounces. How many pounds of cheese did Chu buy?

32. The reporter said yesterday's high temperature was 59°F. What Celsius temperature is that?

33. What is the normal body temperature?

34. During a bad cold, the Smith baby lost 1 pound 5 ounces. After the cold she weighed 13 pounds 3 ounces. How much did she weigh before she caught the cold?

35. Robin wants to mail 2 packages. One weighs 2 pounds 12 ounces, and the other weighs 1 pound 14 ounces. How much do the two packages weigh?

36. Change 35°C to °F.

37. What is the freezing point of water?

38. What is the boiling point of water?

39. Which is warmer, 20°F or 20°C?

Mark the given temperatures on the Fahrenheit thermometer.

40. 103°F

41. 97°F

42. 102.5°F

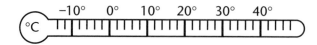

Mark the given temperatures on the Celsius thermometer.

43. 24°C   **45.** 7°C

44. 12°C   **46.** −5°C

At the right is a thermostat for a furnace.

47. At what temperature do you set your thermostat in the winter for a comfortable daytime temperature? Mark the thermostat.

_____

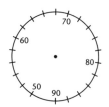

48. At what temperature do you set your thermostat in the winter for a comfortable sleeping temperature? Mark the thermostat.

_____

Mark the given temperatures on the thermostat.

49. 72°F   **51.** 78°F

50. 54°F   **52.** 69°F

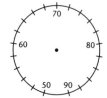

Below is an outdoor Celsius thermometer. It measures how warm the air is. Mark the given temperatures on the thermometer.

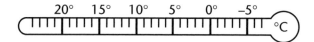

53. 16°C   **54.** 2°C   **55.** 21°C   **56.** −5°C

Answer each question.

57. What Celsius reading is a comfortable outside temperature? _____

58. At what Celsius temperature might it be snowing? _____

List the names of the units.

59. List four basic customary units for measuring length.

_____

60. List three basic customary units for measuring weight.

_____

61. List three basic metric units for measuring length.

_____

62. List three basic metric units for measuring weight.

_____

Write the measurement shown.

63.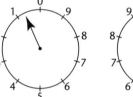

_____

64.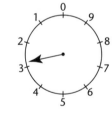

_____

65.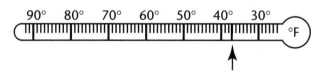

_____

66.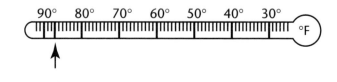

_____

Section 4
Capacity and Volume

The Food Service Professions

People who prepare food to be served in cafeterias, hotels, and restaurants are members of the food services professions. Preparations start early in the morning, with the weighing and buying of fresh vegetables, poultry, fish, and meats. Imagine preparing food for more than 500 people! How much ground beef would you need for a meat sauce? How many onions? How much garlic? What size pots and pans? What seems like a simple meal takes quite a bit of planning when cooking for large numbers of people. Food preparers must be able to convert ounces to pounds, quarts to gallons, teaspoons to tablespoons, and so on.

After buyers purchase the food, kitchen workers perform the cutting, chopping, measuring, counting, and other tasks to prepare ingredients for the various items on the menu. When customers arrive, managers and wait staff take their orders quickly and accurately, and kitchen and bar staff prepare food and drinks. Other workers set up and clear tables, and clean and dry the dishes, glasses, silverware, and pans.

How many things can you name that are measured when you purchase, prepare, and serve food?

Suppose you manage a lunch service in an office building. Here are some of the questions you might ask each day. Ask yourself what you would measure to answer these questions.

- How many people can be served at one time?

- To prepare each item on the menu in large quantities, how much of each ingredient is needed?

- How can you decide which of two large cans of food is less expensive when the cans are different sizes?

- To prepare sauces and soups in large quantities, what sizes of saucepans, kettles, and pots will you need?

Many of the questions addressed by workers in the food service professions involve capacity, volume, and other types of measurements. In these lessons, you will learn about different systems of units for measuring capacity and volume. Also, you will investigate and solve problems related to these units of measurement.

Career Corner

Here is a list of some of the jobs in the food services professions:

chef	buyer	kitchen staff
cook	host or hostess	wait staff
dishwasher	restaurant manager	bus staff

Think of the many other jobs involved in bringing food to people at work, at home, on vacation, or at leisure. Can you name some of those jobs?

Customary Units of Capacity

What units do you use when you want to find out how much something holds? When you use gallons, cups, and teaspoons you are using customary units of capacity. A teaspoon is a very small customary unit of capacity, and a gallon is a large unit. These are the relationships among customary units of capacity.

1 gallon (gal) = 4 quarts (qt)	1 gal = 16 c
1 qt = 2 pints (pt)	1 c = 16 tablespoons (tbsp)
1 pt = 2 cups (c)	1 tbsp = 3 teaspoons (tsp)

Here are some examples of approximate capacity.

dose of
cough medicine
1 tablespoon

serving of coffee
1 cup

container
of ice cream
1 gallon

container
of sour cream
1 pint

container
of milk
1 quart

Name three things that are measured, served, or packaged in the given unit.

1. gallons _____

2. cups _____

3. tablespoons _____

Complete each measurement by writing tsp, tbsp, c, pt, qt, or gal.

4. a bottle of cola   2 _____   8. a dose of medicine $1\frac{1}{2}$ _____

5. a serving of hot tea   1 _____   9. flour to make a cake   3 _____

6. oil for a car   5 _____   10. water to fill a rain barrel 25 _____

7. a jar of peanut butter   1 _____   11. salt in a cookie recipe $\frac{1}{2}$ _____

For each exercise, circle the largest capacity and underline the smallest capacity.

12. oil for brownies, oil for a lawnmower, oil for a car

13. a picnic jug, a swimming pool, a child's wading pool

14. a gallon of milk, a cup of water, a pint of juice

15. a can of peaches, a bottle of ketchup, a bottle of vanilla

16. flour to make cookies, salt to make cookies, oil to make cookies

Here are some common containers used to measure capacity. Some are used to measure very small amounts, and others are used to measure larger amounts. Which of these have you used?

Example 1

Many people have used a measuring cup. Reading the scale is like reading a ruler. The scale is labeled in cups. Each cup is divided into 4 parts, and each mark represents 4 tablespoons. Kai poured rice into the container. How much rice did he pour?

Step 1. Decide which units to use for the measurement.	___ c and ___ tbsp
Step 2. The amount of rice is between 1 and 2 cups.	1 c and ___ tbsp
Step 3. Find the number of tablespoons more than 1 cup.	1 c and 12 tbsp

Answer: Kai poured 1 cup and 12 tablespoons of rice.

Write the number of cups and tablespoons shown in each container.

17.

18.

19.

_____   _____   _____

Estimating Capacity

Here are two tools used to measure capacity. One is used for very small amounts, and the other is used for larger amounts.

This measuring cup can be used for doses of medicine or other small amounts. It uses teaspoons as a basic unit.

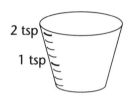

This 2-quart container uses cups as a basic unit.

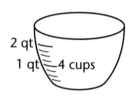

── Example 1 ──────────

To make juice from a can of frozen concentrate, you fill the can four times with water, and add that water to the concentrate. Estimate how many quarts of juice you can make from $\frac{2}{3}$ cup of frozen concentrate. Then mark your estimate on the container.

Step 1. Calculate the total number of cups.

$$\frac{2}{3} \text{ cup} + 4\left(\frac{2}{3}\text{cup}\right) = \frac{2}{3} + \frac{8}{3}$$
$$= \frac{10}{3}$$
$$= 3\frac{1}{3} \text{ cups}$$

Step 2. Estimate the number of quarts in $3\frac{1}{3}$ cups.

$4 \text{ c} = 1 \text{ qt}$, so $3\frac{1}{3} \text{ c} \approx \frac{3}{4} \text{ qt}$

Answer:

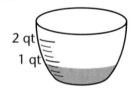

Estimate the capacity of each object. Draw an appropriate measuring container and mark your estimate on the container.

1. a can of evaporated milk

2. a serving of tomato juice

3. amount of water to cook spaghetti

4. carbonated water for a punch recipe

5. a dose of cough syrup

6. amount of coffee in a thermos bottle

7. amount in a giant-size soft drink

8. amount of medicine in an eye dropper

9. water to make a cake mix

10. buttermilk to use in a double batch of muffins

11. amount of canned tomatoes to make chili

12. one serving of soup

Changing Among Customary Units of Capacity

When you use a smaller unit to measure an amount of liquid, the number of units in your measurement will be larger. When you use a larger unit, the number of units in your measurement will be smaller.

Here are the relations among teaspoons, tablespoons, cups, pints, quarts, and gallons.

1 gallon (gal) = 4 quarts (qt)		1 fluid ounce (fl oz) = 2 tbsp	
1 qt = 2 pints (pt)		1 tbsp = 3 tsp	
1 pt = 2 cups (c)		1 c = 8 fl oz	
		1 pt = 16 fl oz	

Example 1

Gail checked the oil in her car. She found her car needed 2 quarts of oil. In her garage, she had three 1-pint containers of oil. Did she have enough oil for her car?

Step 1.	Write the problem.	2 qt = ____ pt
Step 2.	Write the relationship between pints and quarts.	1 qt = 2 pt
Step 3.	Multiply each side of the equation in Step 2 by 2.	2 qt = 4 pt
Answer:	Gail had only 3 pints of oil, so she did not have enough.	

Example 2

Find the number of fluid ounces in 2 cups.

Step 1.	Write the problem.	2 c = ____ fl oz
Step 2.	Write the relationship between cups and fluid ounces.	1 c = 8 fl oz
Step 3.	Multiply both sides of the equation in Step 2 by 2.	2 c = 16 fl oz
Answer:	There are 16 fluid ounces in 2 cups.	

Rewrite each measurement so it uses the given unit.

1. 6 tbsp = ____ tsp

2. 10 gal = ___ pt

3. 3 qt = ___ c

4. 2 pt = ___ fl oz

5. 4 gal = ___ qt

6. 1 c = ___ tbsp

7. 20 qt = ___ pt

8. 12 fl oz = ___ tbsp

9. 25 pt = ___ c

Sometimes a measurement contains a mixed number. Here is an example.

Example 3

Change $5\frac{1}{2}$ gallons to quarts.

Step 1.	Write the problem.	$5\frac{1}{2}$ gal = ___ qt
Step 2.	Write the relationship between gallons and quarts.	1 gal = 4 qt
Step 3.	Multiply both sides of the. equation in Step 2 by $5\frac{1}{2}$.	$5\frac{1}{2}$ gal = 22 qt

Answer: $5\frac{1}{2}$ gal = 22 qt

Rewrite each measurement so it uses the given unit.

10. $1\frac{1}{2}$ c = ___ tbsp

11. $3\frac{5}{8}$ gal = ___ pt

12. $4\frac{1}{3}$ qt = ___ c

13. $2\frac{1}{4}$ tbsp = ___ tsp

14. 2.5 fl oz = ___ tsp

15. $4\frac{3}{4}$ pt = ___ fl oz

16. 6.6 gal = ___ qt

17. $7\frac{1}{2}$ pt = ___ c

In the next two examples, smaller units are replaced with larger units.

Example 4

Tonya plans to make ice cream. Her recipe calls for 6 cups of cream. At the store she finds that the cream is sold in pint containers. How many pints of cream should she buy?

Step 1.	Write the problem.	6 c = ___ pt
Step 2.	Write the relationship between cups and pints.	2 c = 1 pt
Step 3.	Multiply both sides of the equation in Step 2 by 3.	6 c = 3 pt

Answer: She needs to buy 3 pints of cream.

Example 5

Change 9 teaspoons to tablespoons.

Step 1.	Write the problem.	9 tsp = ___ tbsp
Step 2.	Write the relationship between teaspoons and tablespoons.	3 tsp = 1 tbsp
Step 3.	Multiply both sides of the equation in Step 2 by 3.	9 tsp = 3 tbsp
Answer:	Nine teaspoons is the same as three tablespoons.	

Rewrite each measurement so it uses the given unit.

18. 2 c = ___ qt   **21.** 6 qt = ___ gal   **24.** 32 fl oz = ___ pt

19. 32 tbsp = ___ c   **22.** 27 tsp = ___ tbsp   **25.** 160 c = ___ qt

20. 10 c = ___ pt   **23.** 8 pt = ___ gal   **26.** 9 tsp = ___ tbsp

Sometimes you need to use a mixed number for the smaller units. Here is an example.

Example 6

Change 7 cups to pints.

Step 1.	Write the problem.	7 c = ___ pt
Step 2.	Write the relationship between cups and pints.	2 c = 1 pt
Step 3.	Multiply both sides of the equation in Step 2 by $3\frac{1}{2}$.	7 c = $3\frac{1}{2}$ pt
Answer:	7 c = $3\frac{1}{2}$ pt	

Rewrite each measurement so it uses the given unit.

27. 8 fl oz = ___ pt   **30.** 4 tsp = ___ tbsp   **33.** 7 c = ___ qt

28. 22 qt = ___ gal   **31.** 30 fl oz = ___ pt   **34.** 12 fl oz = ___ c

29. 20 tsp = ___ tbsp   **32.** 8 pt = ___ gal   **35.** 25 tbsp = ___ c

Solve each problem.

36. Mary Jo needed to give her baby half a teaspoon of medicine. The easiest way to feed her baby was by using a baby bottle. How many fluid ounces of medicine should be in her baby's bottle?

37. Jose displayed 120 quarts of strawberries at the fruit store. How many gallons did he display?

38. Tom has an 8-quart coffee maker. How many cups of coffee will it make?

39. A soup recipe calls for 3 pints of cream. How many cups of cream does the recipe use?

40. The soup recipe also calls for 3 tablespoons of salt. How many teaspoons of salt does the recipe call for?

41. The soap dispenser for Ann's dishwasher holds 2 fluid ounces of liquid soap. How many tablespoons of soap will the dishwasher hold?

42. Jack used a quart container to fill a 10-gallon pot with soil so he could plant a large flowering shrub. How many quarts of soil did it take to fill the pot?

43. If Jack fills the pot only half full, how many pints of soil will the pot contain?

44. Betty and Brad prepare cans of vegetables from their garden. They canned 20 pints of tomatoes. How many quarts did they can?

45. Wendy was told to drink 32 fluid ounces of water every day. How many cups of water does she need to drink daily?

46. David feeds his dog 1 cup of food every day. After the fourth day, how many pints of food has David fed his dog?

47. A recipe for hot chili calls for 6 teaspoons of chili powder. How many tablespoons of chili powder are required for this recipe?

Estimating with Metric Units of Capacity

In the metric system there are three basic units of capacity. They are **liters**, **milliliters**, and **kiloliters**. Here are the relationships among these units.

$$1 \text{ liter (L)} = 1{,}000 \text{ milliliters (mL)}$$
$$1 \text{ kiloliter (kL)} = 1{,}000 \text{ L}$$

In the metric system of measurement, the prefix "milli-" means $\frac{1}{1{,}000}$ or one-thousandth. The prefix "kilo-" means 1,000. Small amounts are measured in milliliters, and large amounts are measured in kiloliters.

10 mL

10 milliliters
a dose of medicine

4 liters
a watering can

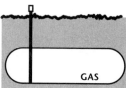

GAS

2 kiloliters
an underground gas tank

Name three things that are measured in the given unit.

1. milliliters

_____

2. liters

_____

3. kiloliters

_____

Circle the most appropriate unit for measuring the capacity of each item.

4. a sample of blood   mL   L   kL

5. water in a swimming pool   mL   L   kL

6. oil for a can   mL   L   kL

7. amount a tanker holds   mL   L   kL

8. amount a picnic jug holds   mL   L   kL

9. water for a hot tub   mL   L   kL

10. shot of antibiotic   mL   L   kL

11. a keg of beer   mL   L   kL

12. oil to fry an egg   mL   L   kL

Underline the better estimate for the capacity of each object.

13. juice in an orange   16 L   16 mL

14. coffee in a thermos   1 L   1 kL

15. water in a bathtub   28 L   28 kL

16. bleach for 16 loads of laundry   2 kL   2 L

17. salt in a cake mix   3 L   3 mL

18. oil for a car   2.5 kL   2.5 L

19. water in a kitchen sink   4 mL   4 L

20. water to make fruit punch   2 mL   2 L

21. oil from an oil well   6 kL   6 mL

22. water used to take a shower   16 L   16 kL

23. lava from a volcano   206 L   206 kL

24. punch at a wedding reception   16 L   16 kL

25. juice in a can of pineapple   240 L   240 mL

26. blood for a cholesterol test   25 mL   25 L

27. water in a fountain   14 mL   14 kL

Answer each exercise.

28. Circle the smallest unit.   1 kL   1 L   1 mL

29. Circle the largest unit.   1 kL   1 L   1 mL

30. What are the three basic metric units for capacity?

_____

Changing Among Milliliters, Liters, and Kiloliters

Here is a review of the relationships among milliliters, liters, and kiloliters. They are the three basic metric units for capacity.

$$1 \text{ L} = 1,000 \text{ mL}$$
$$1 \text{ kL} = 1,000 \text{ L}$$

Example 1

Joe and Jana were dehydrated after playing ball in the hot sun all day. The nurse gave each of them $\frac{1}{4}$ liter of fluid. How many milliliters of fluid did each child receive?

Step 1.	Write the problem.	$\frac{1}{4}$ L = ____ mL
Step 2.	Write the relationship between milliliters and liters.	1 L = 1,000 mL
Step 3.	Multiply each side of the equation in Step 2 by $\frac{1}{4}$.	$\frac{1}{4}$ L = 250 mL

Answer: Each child received 250 milliliters of fluids.

Example 2

Change 2 liters to kiloliters.

Step 1.	Write the problem.	2 L = ____ kL
Step 2.	Write the relationship between liters and kiloliters.	1,000 L = 1 kL
Step 3.	Divide both sides of the equation in Step 2 by 500.	$2 \text{ L} = \frac{1}{500} \text{ kL}$ = 0.002 kL

Answer: 2 liters = 0.002 kiloliters

Rewrite each measurement so it uses the given unit.

1. 4 L = ____ mL

2. 3 L = ____ kL

3. 600 mL = ____ L

4. 10 L = ____ kL

5. 6 L = ____ mL

6. 5 kL = ____ L

Sometimes the given metric measurement has a decimal point. Here is an example.

___ Example 3 _____

Change 4.5 liters to milliliters.

Step 1.	Write the problem.	4.5 L = ____ mL
Step 2.	Write the relationship between liters and milliliters.	1 L = 1,000 mL
Step 3.	Multiply both sides of the 4 equation in Step 2 by 4.5.	4.5 L = 4,500 mL

Answer: 4.5 liters = 4,500 milliliters

Rewrite each measurement so it uses the given unit.

7. 1.5 L = ____ mL   10. 3.2 L = ____ kL

8. 2,400 mL = ____ L   11. 3.2 kL = ____ L

9. 5.5 L = ____ mL   12. 0.05 kL = ____ L

Sometimes you need to use a decimal point with a larger unit. Here is an example.

___ Example 4 _____

Change 350 milliliters to liters.

Step 1.	Write the problem.	350 mL = ___ L
Step 2.	Write the relationship between milliliters and liters.	1,000 mL = 1 L
Step 3.	Set up a proportion. Use "?" to represent the unknown amount.	$\dfrac{350}{1,000} = \dfrac{?}{1}$
Step 4.	Find the cross-products and solve for the unknown number.	$? \times 1,000 = 350$ $? = \dfrac{350}{1,000} = 0.35$

Answer: 350 milliliters = 0.35 liters

Here are two more examples of changing a measurement so it uses a larger unit.

Example 5

Patrice works in a nursing home. She makes sure her patients drink at least 2,000 milliliters of water every day. How many liters of water should each patient drink?

Step 1. Write the problem. $\qquad$ 2,000 mL = _____ L

Step 2. Write the relationship $\qquad$ 1,000 mL = 1 L
between milliliters and liters.

Step 3. Multiply each side of the $\qquad$ 2,000 mL = 2 L
equation in Step 2 by 2.

Answer: Each patient should drink 2 liters of water every day.

Example 6

Change 400 milliliters to kiloliters.

Step 1. Find the relationship between $\qquad$ 1 kL = 1,000 L
milliliters and kiloliters. Each kiloliter $\qquad$ 1 L = 1,000 mL
is 1,000 liters, and each liter is
1,000 milliliters. $\qquad$ So 1 kL = 1,000,000 mL.
In each kiloliter there are
1,000 × 1,000 = 1,000,000 milliliters.

Step 2. Decide how to convert milliliters $\qquad$ To change kL to mL, you
to kiloliters. $\qquad$ multiply by 1,000,000.
So to change from mL
to kL, you divide by 1,000,000.

Step 3. Convert 400 milliliters to kiloliters. $\qquad$ $\frac{400}{1,000,000} = \frac{4}{10,000} = 0.0004$

Answer: 400 milliliters = 0.0004 kiloliters

Rewrite each measurement so it uses the given unit.

13. 10,000 mL = _____ L

17. 100,000 mL = _____ L

14. 600 mL = ___ kL

18. 800 mL = ___ kL

15. 150 mL = ___ L

19. 6,200 mL = _____ L

16. 780 mL = ___ kL

20. 9,225 mL = _____ kL

Adding and Subtracting with Capacity Measurements

Sometimes you need to add or subtract measurements of capacity.

Example 1

Sam used 3 gallons and 3 quarts of paint to paint his house. He used 2 gallons and 3 quarts to paint his garage. How much paint did he use in all?

Step 1. Line up the measurements, putting like units under like units.

$$\begin{array}{r} 3 \text{ gal } 3 \text{ qt} \\ + 2 \text{ gal } 3 \text{ qt} \\ \hline \end{array}$$

Step 2. Add the quarts and add the gallons.

$$\begin{array}{r} 3 \text{ gal } 3 \text{ qt} \\ + 2 \text{ gal } 3 \text{ qt} \\ \hline 5 \text{ gal } 6 \text{ qt} \end{array}$$

Step 3. Think of 6 quarts as 4 quarts + 2 quarts. So 6 quarts = 1 gallon 2 quarts. Rewrite the sum.

$$\begin{aligned} 5 \text{ gal} + 6 \text{ qt} &= 5 \text{ gal} + 4 \text{ qt} + 2 \text{ qt} \\ &= 5 \text{ gal} + 1 \text{ gal} + 2 \text{ qt} \\ &= 6 \text{ gal } 2 \text{ qt} \end{aligned}$$

Answer: He used 6 gallons and 2 quarts of paint.

Example 2

Subtract 3 pints and $1\frac{1}{2}$ cups from 5 pint and 1 cup.

Step 1. Line up the measurements, putting like units under like units.

$$\begin{array}{r} 5 \text{ pt } 1 \text{ c} \\ - 3 \text{ pt } 1\frac{1}{2} \text{ c} \\ \hline \end{array}$$

Step 2. "Borrow" by changing 5 pints to 4 pints + 2 cups. So Rewrite the top number as 4 pints + 3 cups.

$$\begin{array}{r} 4 \text{ pt } 3 \text{ c} \\ - 3 \text{ pt } 1\frac{1}{2} \text{ c} \\ \hline \end{array}$$

Step 3. Subtract the pints and subtract the cups.

$$\begin{array}{r} 4 \text{ pt } 3 \text{ c} \\ - 3 \text{ pt } 1\frac{1}{2} \text{ c} \\ \hline \end{array}$$

Answer: The difference is 1 pint $1\frac{1}{2}$ cups.

$$\begin{array}{r} 4 \text{ pt } 3 \text{ c} \\ - 3 \text{ pt } 1\frac{1}{2} \text{ c} \\ \hline 1 \text{ pt } 1\frac{1}{2} \text{ c} \end{array}$$

Add or subtract.

1. $\begin{array}{r} 6 \text{ gal } 3 \text{ qt} \\ + 2 \text{ gal } 2 \text{ qt} \\ \hline \end{array}$

4. $\begin{array}{r} 2 \text{ gal } 1 \text{ qt} \\ + 3 \text{ gal } 1 \text{ qt} \\ \hline \end{array}$

7. $\begin{array}{r} 6 \text{ pt } 0 \text{ c} \\ - 3 \text{ pt } 1 \text{ c} \\ \hline \end{array}$

10. $\begin{array}{r} 2 \text{ pt } 1 \text{ c} \\ + 6 \text{ pt } 1 \text{ c} \\ \hline \end{array}$

2. $\begin{array}{r} 6 \text{ c } 3 \text{ tbsp} \\ - 2 \text{ c } 6 \text{ tbsp} \\ \hline \end{array}$

5. $\begin{array}{r} 2 \text{ qt } 3 \text{ c} \\ + 6 \text{ qt } 2 \text{ c} \\ \hline \end{array}$

8. $\begin{array}{r} 2 \text{ qt } 3 \text{ pt} \\ + 6 \text{ qt } 2 \text{ pt} \\ \hline \end{array}$

11. $\begin{array}{r} 4 \text{ qt } 2 \text{ c} \\ - 1 \text{ qt } 3 \text{ c} \\ \hline \end{array}$

3. $\begin{array}{r} 12 \text{ gal } 13 \text{ c} \\ + 4 \text{ gal } 8 \text{ c} \\ \hline \end{array}$

6. $\begin{array}{r} 7 \text{ pt } 1 \text{ c} \\ - 3 \text{ pt } 2 \text{ c} \\ \hline \end{array}$

9. $\begin{array}{r} 6 \text{ qt } 1 \text{ c} \\ - 4 \text{ qt } 3 \text{ c} \\ \hline \end{array}$

12. $\begin{array}{r} 4 \text{ gal } 0 \text{ qt } 0 \text{ pt} \\ - \quad\quad 2 \text{ qt } 1 \text{ pt} \\ \hline \end{array}$

To add or subtract metric measurements, make sure the measurements use the same unit.

—— Example 3 ——

Find the sum and the difference of 5.2 L and 1,350 mL.

Step 1.	Rewrite one of the measurements so both use the same units. This example uses milliliters.	5.2 L = ___ mL 1 L = 1,000 mL 5.2 L = 5,200 mL	

Step 2. Line up the two measurements. Then add or subtract.

	Add	Subtract
	5,200 mL	5,200 mL
	+ 1,350 mL	− 1,350 mL
	6,550 mL	3,850 mL

Answer: The sum is 6,550 milliliters. The difference is 3,850 milliliters.

Find each sum or difference.

13. 35 L + 2 kL

16. 8.65 L – 1,400 mL

14. 3 L – 2,540 mL

17. 2.35 L + 450 mL

15. 7.2 L + 3,500 L

18. 2.6 kL – 1,750 L

Solve each problem.

19. If Juan starts with 2 gallons 3 quarts and adds 2 gallons 2 quarts, how much total liquid does he have?

20. Sandra took 750 milliliters out of a full 2.5-liter container of hydrogen peroxide. How much was left?

21. Sam added 2.5 liters of punch to a bowl that contained $6\frac{1}{2}$ liters of punch. How much punch was in the bowl?

22. A soup recipe made 1 gallon 5 cups of soup. If Jarod served 16 cups of soup at a dinner, how much soup was left?

Multiplying and Dividing with Capacity Measurements

Sometimes you need to multiply or divide measurements of capacity. For example, to find the total capacity of several same-size containers, you can multiply. To find out how many times you can fill a small container from a large one, you can divide.

Example 1

Digna owns her own auto shop, and she buys car oil in 15-gallon drums. If the average car uses $1\frac{1}{2}$ gallons of oil, how many cars can she service with one 15-gallon container of oil?

Step 1. Decide which math operation to use.

She wants to know how many times a large container (the 15-gallon drum) can be used to fill smaller containers (the cars). The operation is division.

Step 2. Write the problem.

$$15 \div 1\frac{1}{2} = 15 \div \frac{3}{2}$$

Step 3. Divide 15 by $\frac{3}{2}$. That is the same as multiplying 15 by $\frac{2}{3}$.

$$= \frac{15}{1} \times \frac{2}{3}$$

Step 4. Multiply the two fractions.

$$= \frac{15 \times 2}{1 \times 3} = \frac{30}{3} = 10$$

Answer: Digna can service 10 cars with one 15-gallon drum of oil.

Example 2

Multiply 6 quarts 2 cups by 5.

Step 1. Write the problem.

$$\begin{array}{r} 6 \text{ qt } 2 \text{ c} \\ \times \quad\quad 5 \\ \hline \end{array}$$

Step 2. Multiply 5 times the number of cups and multiply 5 times the number of quarts.

$$\begin{array}{r} 6 \text{ qt } 2 \text{ c} \\ \times \quad\quad 5 \\ \hline 30 \text{ qt } 10 \text{ c} \end{array}$$

Step 3. One quart is 4 cups, so replace 10 cups with 2 quarts and 2 cups.

$$30 \text{ qt } 10 \text{ c} = 30 \text{ qt} + 2 \text{ qt} + 2 \text{ c}$$
$$= 32 \text{ qt } 2 \text{ c}$$

Answer: The product is 32 quarts 2 cups.

— **Example 3** —

Divide 4 liters and 800 milliliters by 12.

Step 1.	Use the same unit for both measurements. Using liters, 800 milliliters = 0.8 liters.	4 L + 800 mL = 4 L + 0.8 L = 4.8 L
Step 2.	Write the problem.	$\frac{4.8}{12} = ?$
Step 3.	Divide.	$\frac{4.8}{12} = 0.4$

Answer: The quotient is 0.4 liters.

Multiply or divide.

1. 2 pt 1 c × 6

2. 4 gal 2 qt ÷ 2

3. 5 L 250 mL ÷ 5

4. 3 kL 500 L ÷ 2

5. 3 gal 3 qt × 6

6. 3 qt 1 pt × 10

7. 6 gal 2 qt ÷ 4

8. 6.4 L × 8

9. 2.7 kL ÷ 3

10. 16 L 200 mL ÷ 4

11. 6 qt 4 c ÷ 4

12. 2 gal 3 pt × 6

13. 12 qt 1 c × 4

14. 6 c 4 tbsp × 3

15. 6 kL 500 mL ÷ 25

Solve each problem.

16. Tami bought 5 cans of evaporated milk. Each has a capacity of 354 milliliters. How many milliliters did she buy? How many liters?

17. Harold brought 2 gallons and 6 cups of apple cider to a party. If there were 11 people at the party, how many cups could each person have?

18. Tru is making a double batch of chicken soup. If the recipe calls for 1 quart 3 cups of broth, how much broth does he need?

19. Sami needs 80 milliliters of liquid to run a lab test. She wants to run the test twelve times. How much liquid will she use?

20. The Folden family has $2\frac{1}{2}$ quarts of liquid detergent. If they use $\frac{1}{2}$ cup of detergent in each load of wash, how many loads of wash can they do?

Comparing Customary and Metric Units of Capacity

Sometimes a measurement of capacity is listed in both metric units and customary units. For example, the wrapper of a large soda container lists the capacity as "2 liters (67.6 fl oz)".

At the right are some comparisons between metric and customary units.

1 L ≈ 1.1 qt (so 1 L > 1 qt)
1 mL ≈ 2 drops
1 kL = 1,000 L ≈ 275 gal

Example

There were 2 displays at the grocery store. One had a liter bottle of juice for 75¢. The other had a quart bottle for 75¢. Which was the better buy?

Step 1. Write the two units   1 L   1 qt
Step 2. Decide which is larger.   1 L > 1 qt
Step 3. Compare the prices.   The prices are the same.
Answer: One liter is larger than one quart, so one liter for 75¢ is a better buy than one quart for 75¢.

Circle the larger measurement.

1. 1 qt or 1 mL

2. 1 gal or 1 kL

3. 1 L or 1 c

4. 1 mL or 1 fl oz

5. 1 pt or 1 L

6. 1 gal or 1 L

7. 1 tsp or 1 mL

8. 1 qt or 1 gal

9. 1 kL or 1 L

10. 1 c or 1 pt

11. 1 qt or 1 kL

12. 1 fl oz or 1 tbsp

Circle the more appropriate measurement for the capacity of each object.

13. can of evaporated milk   400 mL or 400 fl oz

14. water in a swimming pool   1,000 kL or 1,000 qt

15. oil in a small truck   6 qt or 6 kL

16. glass of juice   1 L or 1 c

17. gas in a full tank of a sports car   15 L or 15 gal

18. container of bleach in a laundry room   1 gal or 1 kL

19. oil to make brownies   6 fl oz or 6 mL

Focus on Geometry: Volume

Volume (*V*) is the amount of space inside a three-dimensional figure. Volume is measured in cubic units. For a three-dimensional figure shaped like a box, you can use the formula $V = \ell \times w \times h$ to calculate volume.

Example 1

Carla's boss asked her for the volume of a packing box that is 18 inches by 12 inches by 15 inches. What should she do to calculate the volume? What is the volume of the box?

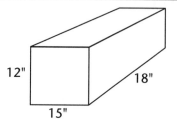

Step 1. Write the formula for volume. $V = \ell \times w \times h$

Step 2. Substitute the values of ℓ, w, and h into the formula. $V = 18 \times 12 \times 15$

Step 3. Multiply the numbers. $V = 3{,}240$

Answer: She should multiply the three measurements. The volume of the box is 3,240 cubic inches.

Find the volume of each box. Use the formula $V = \ell \times w \times h$.

1.

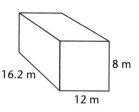

$V = \underline{\hspace{2cm}}$

3.

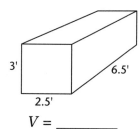

$V = \underline{\hspace{2cm}}$

2.

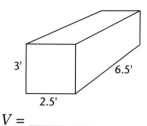

$V = \underline{\hspace{2cm}}$

4.

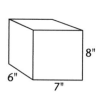

$V = \underline{\hspace{2cm}}$

5. Name three situations in which you might want to calculate volume.

Sometimes the dimensions are not given in the same units. Before you can use the volume formula, all three dimensions must be given in the same unit.

Example 2

Find the volume of a box 4 feet long, 2 feet wide, and 9 inches tall.

Step 1. Use a single unit for all three measurements. To use feet as the unit, change 9 inches to feet.

$$9 \text{ in.} = \frac{9}{12} \text{ ft}$$
$$= \frac{3}{4} \text{ ft}$$

Step 2. Write the formula.

$$V = \ell \times w \times h$$

Step 3. Substitute the values of ℓ, w, and h into the formula.

$$V = 4 \times 2 \times \frac{3}{4}$$

Step 4. Multiply.

$$V = 6$$

Answer: The volume is 6 cubic feet.

Find the volume of each box.

6. $\ell = 10$ ft
   $w = 1$ yd
   $h = 5$ ft
   $V =$ _____

7. $\ell = 6$ m
   $w = 500$ mm
   $h = 2$ m
   $V =$ _____

8. $\ell = 3$ yd
   $w = 1$ ft
   $h = 2$ yd
   $V =$ _____

9. $\ell = 2$ in.
   $w = 6$ in.
   $h = 18$ in.
   $V =$ _____

10. $\ell = 4$ yd
    $w = 36$ in.
    $h = 6$ ft
    $V =$ _____

11. $\ell = 350$ mm
    $w = 2$ m
    $h = 0.05$ m
    $V =$ _____

A **cube** is a three-dimensional figure that has squares as all the faces of the figure. If the length of each side of each square is *s*, then a formula to find the volume of a cube is $V = s^3$.

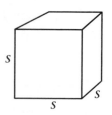

s = length of one side
$V = s \times s \times s$ or s^3

12. Name some common objects that are cubes.

_____

Find the volume of each cube.

13.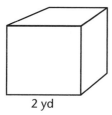

2 yd

$V = $ _____

15.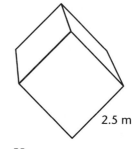

2.5 m

$V = $ _____

14.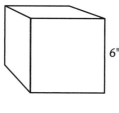

6"

$V = $ _____

16.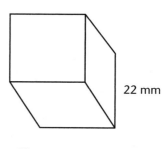

22 mm

$V = $ _____

Solve each problem.

17. What is the volume of a cube whose sides are 15 inches?

18. What is the volume of a rectangular box that is 6 inches by 8 inches by 3 inches?

19. A sugar cube is about a half-inch on each side. What is the volume of the sugar cube?

20. What is the volume of a cube whose sides are 6.4 meters?

If you know the volume of a box and measurements for two of its dimensions, you can use the formula $V = \ell \times w \times h$ to find the measurement for the third dimension.

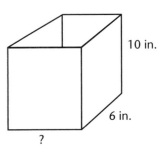

Example 3

The volume of a box is 480 cubic inches. If the height is 10 inches and the width is 6 inches, what is the length of the box?

Step 1.	Write the formula.	$V = \ell \times w \times h$
Step 2.	Substitute the known values into the formula. Don't forget that the value for V is known.	$480 = \ell \times 6 \times 10$
Step 3.	Simplify the expression.	$480 = 60 \times \ell$
Step 4.	Divide both sides of the equation in Step 3 by 60.	$8 = \ell$
Answer:	The length of the box is 8 inches.	

Find the missing dimension for each box.

21. $V = 48$ cu ft

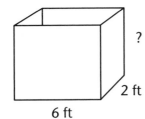

22. $V = 1,200$ cu m

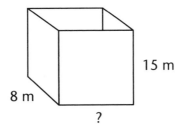

23. $V = 800$ cu in.

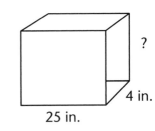

Solve each problem.

24. The volume of a box is 1,200 cubic inches. The width is 6 inches and the height is 10 inches. What is the length of the box?

25. The volume of a cube is 1,000 cubic inches. How long is each side?

26. The volume of a cereal box is 576 cubic inches. If the height of the box is 1 foot and the width 4 inches, what is the length of the box?

27. If the area of one side of a cube is 64 square inches, what is its volume?

28. Ray is digging a trench for an underground sprinkler. If the trench is 6 inches deep, 2 feet wide, and has a volume of 3 cubic feet, how long is the trench?

Focus on Geometry: Volume of Cylinders and Cones

A **cylinder** is a can-shaped container. The top
and bottom, which are circles, are called the
bases of the cylinder. To calculate the volume
of a cylinder, you can multiply the height times
the area of the base.

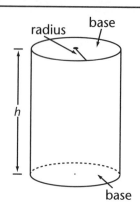

Volume of a cylinder = area of base × height
$V = B \times h$ where B is the area of the base
$V = \pi r^2 h$ where $\pi \approx 3.14$ and r is the radius of the base

Example 1

Freeda wants to fill a large container with dirt so she can plant some tomatoes. The
container is a cylinder that is 2 feet tall and has a radius of 3 feet. What is the volume
of the container?

Step 1.   Write the formula. $V = \pi r^2 h$
Step 2.   Substitute the values for π, r,
             and h into the formula. $V = (3.14)(3^2)(2)$
Step 3.   Evaluate 3^2. $3^2 = 3 \times 3 = 9$
Step 4.   Multiply. $V = (3.14)(9)(2)$
                                       $= 56.52$

Answer: The volume of the planter is about 56.5 cubic feet.

Find the volume of each cylinder.

1.

2'

12'

$V = $ _____

3.

8 m

25 m

$V = $ _____

5.

2.5 cm

7.5 cm

$V = $ _____

2.

20 cm

6 cm

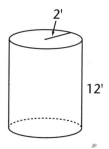

$V = $ _____

4.

10 in.

4 in.

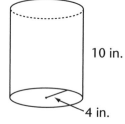

$V = $ _____

6.

15 ft

4 ft

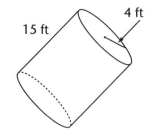

$V = $ _____

Sometimes the dimensions for a cylinder are not given in the same units. Before you can use the volume formula, all the dimensions must use the same unit.

Example 2

Find the volume of a cylinder that is 4 feet high and has a radius of 18 inches.

Step 1. Use a single unit for all the measurements. To use feet as the unit, change 18 inches to feet.

18 in. $= \frac{18}{12}$ ft
$= 1\frac{1}{2}$ ft

Step 2. Write the formula.

$V = \pi r^2 h$

Step 3. Substitute the values for π, r, and h into the formula.

$V = (3.14)(\frac{3}{2})(\frac{3}{2})(4)$

Step 4. Multiply the numbers.

$V = 28.26$ cu ft

Answer: The volume of the cylinder is 28.26 cubic feet.

Find the volume of each cylinder.

7.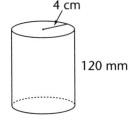

$V = _____$

8.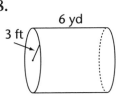

$V = _____$

9.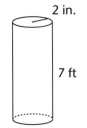

$V = _____$

10.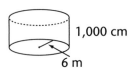

$V = _____$

If you are given the diameter of a cylinder, you can find the radius because the diameter is twice the radius.

Example 3

Find the volume of a cylinder that is 10 feet high and has a diameter of 6 feet.

Step 1. Find the value of the radius.

$r = \frac{1}{2}d = (\frac{1}{2})(6) = 3$

Step 2. Write the formula.

$V = \pi r^2 h$

Step 3. Substitute the values of π, r, and h into the formula. Then multiply.

$V = (3.14)(3)(3)10$
$= 282.6$

Answer: The volume is 282.6 cubic feet.

Find the volume of each cylinder.

11.

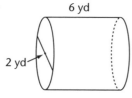

6 yd

2 yd

$V = $ _____

13.

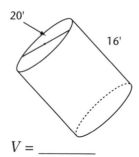

20'

16'

$V = $ _____

12.

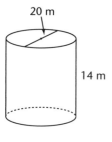

20 m

14 m

$V = $ _____

14.

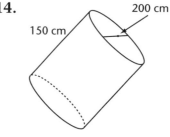

150 cm

200 cm

$V = $ _____

A **cone** is a container that has a circular **base** at one end and a point at the other end. The **height** of a cone is the distance from the center of the circular base to the point.

Here is the formula for the volume of a cone:

$V = \frac{1}{3}\pi r^2 h$ where $\pi \approx 3.14$,
      r is the radius of the base,
      and h is the height.

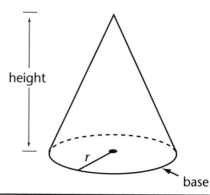

height

r

base

— Example 4 ————————————————

Find the volume of a cone with
height 6 inches and radius 2 inches.

Step 1.   Write the formula.

$V = \frac{1}{3}\pi r^2 h$

Step 2.   Substitute values for π, r,
             and h into the formula.

$V = \frac{1}{3}(3.14)(2^2)(6)$

$= \frac{1}{3}(3.14)(2 \times 2)(6)$

2 in.

6 in.

Step 3.   Multiply.

$V = 25.12$

Answer: The volume of the cone is 25.12 cubic feet.

Find the volume of each cone.

15.

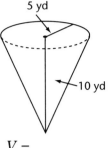

5 yd

10 yd

V = _____

18.

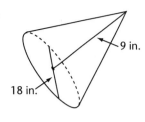

9 in.

18 in.

V = _____

16.

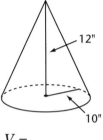

12"

10"

V = _____

19.

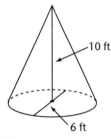

10 ft

6 ft

V = _____

17.

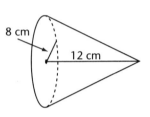

8 cm

12 cm

V = _____

20.

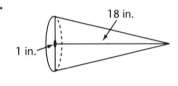

18 in.

1 in.

V = _____

Solve each problem.

21. A cylindrical tank at a gas station is 30 feet long and has a diameter of 18 feet. What is the volume of the tank? If a cubic foot is about 7.5 gallons, approximately how many gallons of gas does the tank hold?

22. The cylindrical gas tank is 6 feet long and has a diameter of 5 feet. What is the volume of the tank?

23. A cylinder is 10 inches tall. If the volume of the cylinder is 785 cubic inches, what is the radius of the cylinder?

24. If the volume of a cone with a 6-centimeter diameter is 282.6 cubic centimeters, what is the height of the cone?

25. What is the volume of a cone that is 6 inches tall and has a radius of 4 inches?

Application: Oil and Coolant

Truck	Amount of Oil Needed for Oil Change (Quarts)	
	New Filter	**No Filter**
VG 30E 4 × 2	$4\frac{1}{8}$	$3\frac{7}{8}$
VG 30E 4 × 4	$3\frac{5}{8}$	$3\frac{1}{8}$
KA 24E 4 × 2	$4\frac{1}{8}$	$3\frac{3}{4}$
KA 24E 4 × 4	$3\frac{1}{2}$	$3\frac{1}{8}$

Truck	Amount of Coolant Needed	
	Liters	**Quarts**
VG 30E 4 × 2	10.7	$11\frac{3}{4}$
VG 30E 4 × 4	11.7	$12\frac{7}{8}$
KA 24E 4 × 2	8.1	$7\frac{3}{8}$
KA 24E 4 × 4	9.0	$9\frac{7}{8}$

1. Replacing the oil and oil filter in the VG 30E 4 × 2 takes $4\frac{1}{8}$ quarts of oil; replacing the oil without replacing the oil filter takes $3\frac{7}{8}$ quarts of oil. What is the difference in the amount of oil needed?

2. If Ray buys a Pathfinder VG 30E 4 × 4, how many quarts of oil will he need to replace all the oil and replace the oil filter?

3. Replacing the oil without replacing the oil filter for the KA 24E 4 × 2 requires $3\frac{3}{4}$ quarts of oil. For the KA 24E 4 × 4, replacing the oil without replacing the oil filter requires $3\frac{1}{8}$ quarts of oil. What is the total for those two amounts of oil?

4. How many liters of coolant (antifreeze) does the VG 30E 4 × 2 hold?

5. The KA 24E 4 × 2 holds 8.1 liters of coolant. What is the difference between this amount and the amount for the truck in Exercise 4?

6. Before adding antifreeze coolant to your radiator, it needs to be mixed in a 50-50 ratio with water. How many liters of water and antifreeze do you need for the VG 30E 4 × 4? How many liters of water and antifreeze do you need for the KA 24E 4 × 4?

7. Change each of the coolant measures below from quarts to pints.

Truck	Amount of Coolant Needed	
	Quarts	**Pints**
VG 30E 4 × 2	$11\frac{3}{4}$	____?____
VG 30E 4 × 4	$12\frac{7}{8}$	____?____
KA 24E 4 × 2	$7\frac{3}{8}$	____?____
KA 24E 4 × 4	$9\frac{7}{8}$	____?____

Application: Recipes

1. Cheri is having a bridal shower for a friend. She wants to make strawberry punch for 25 people. At the store, the lemonade cans contain 12 ounces of frozen lemonade. How many cans should she buy?

2. How many ounces of frozen lemonade will Cheri have left over after she prepares the punch?

3. The strawberry punch recipe calls for seven 16-ounce cans of water. How many cups of water is this?

4. Margaret plans to make baked chocolate pudding for the bridal shower. The recipe calls for $\frac{1}{4}$ cup shortening and 1 cup flour. How many times will she need to fill a quarter-cup measuring container to measure out the flour? How many times will she fill the quarter-cup container to measure out the shortening?

5. The baked chocolate pudding recipe calls for 3 tablespoons of cocoa. Using a half-teaspoon measuring spoon, how many times will she need to fill it to measure out 3 tablespoons?

6. Margaret is going to double the recipe for baked chocolate pudding. Write a list of the ingredients and their amounts that Margaret will need.

7. For another party, Cheri is going to make strawberry punch using half of the recipe. Write a list of ingredients and their amounts for half the strawberry punch recipe.

Strawberry Punch

one 16-oz can frozen orange juice

one 16-oz can frozen lemonade

one 16-oz pkg. frozen strawberries

one 32-oz bottle ginger ale

seven 16-oz cans water

Mix all ingredients in punch bowl and top with fruit slices and ice rings. Serves approximately 25.

Baked Chocolate Pudding

$\frac{1}{4}$ c shortening	Combine shortening, salt, cinnamon and sugar. Cream thoroughly. Sift flour, baking powder, baking soda, and cocoa together. Add to creamed mixture alternately with milk, blending well after each addition. Add nuts. Drop batter by teaspoon on the hot syrup. Bake at 350° for 45 minutes.
1 c sifted flour	
$\frac{1}{2}$ tsp salt	
2 tsp baking powder	
1 tsp cinnamon	
$\frac{1}{2}$ tsp baking soda	
$\frac{1}{4}$ c sugar	
$\frac{2}{3}$ c milk	
3 tbsp cocoa	
$\frac{1}{2}$ c chopped nuts	

Application: Ice, Water, Gas

1. Jeff has a job with a company that delivers ice. He delivers 25-pound cubes of ice to various restaurants in town. Each side of each cube is 300 millimeters. What is the volume of one cube of ice?

2. Jeff also delivers bags of ice for ice machines in convenience stores. Each bag is $1\frac{1}{2}$ feet long, $\frac{1}{2}$ foot wide, and 8 inches across. Find the volume of a bag of ice.

3. On a winter day, water dripped from a branch of a tree and formed a cone of ice. The height of the cone is 18 inches and the radius of the base of the cone is $\frac{1}{2}$ foot. What is the volume of the ice cone?

4. Gabrielle has a new bathtub. The bathtub is 60 inches long, 31 inches wide, and 20 inches high. If she fills the tub full, what is the volume of water in the tub?

5. When Gabrielle bathes her baby, she fills the bathtub $\frac{1}{4}$ full. What is the volume of the water in the tub when she bathes her baby?

6. As a special treat for his birthday, Marco was given a chocolate ice cream cone. The ice cream parlor uses cones that have a height of 5 inches and a base with diameter of 2 inches. Find the volume of the cone.

7. As part of Marco's birthday celebration, he and his friends were driven to a field where a hot air balloon waited. The part of the balloon which carried the passengers, called a **gondola**, was a box-shaped basket that was 6 feet long, 5 feet wide, and 4 feet tall. Find the volume of the gondola.

8. When the hot air balloon is filled, it takes the shape of a ball. That shape is called a **sphere**. You can use the formula $V = \frac{4}{3}\pi r^3$ to find the volume of a sphere, where $\pi \approx 3.14$, r is the radius of the balloon, and r^3 means $r \times r \times r$. Find the volume of the hot air balloon with a radius of 4.5 meters.

9. Theresa makes juice as part of her job preparing breakfast at a restaurant. She uses a frozen concentrate that comes in a can shaped like a cylinder. The can is eight inches tall and has a diameter of 5 inches. To make the juice, she adds three containers of water to the frozen concentrate. What is the volume of the water that she adds to the frozen concentrate?

10. Theresa had a strawberry malt for a treat. As she drank it, she practiced her math. She measured the malt container and found it to be a cylinder 7 inches tall with a radius of 2 inches. What is the volume of the cylinder?

Application: Heating and Air Conditioning

In order to design a heating and air conditioning system for a building, you have to know the interior volume of the building. For any building, the interior volume can be calculated by multiplying the number of square feet of floor area times the height of the ceiling.

$$V = \text{floor area} \times \text{ceiling height}$$

1. A room with floor area of 500 square feet and an 8-foot-high ceiling has what volume?

2. An apartment with a floor area of 1,800 square feet and an 8-foot-high ceiling has what volume?

3. A house has a volume of 16,000 cubic feet and has a floor area of 2,000 square feet. What is the height of the ceiling?

4. An office has a volume of 30,000 cubic feet. If the ceiling is 10 feet high, what is the floor area of the office?

5. What is the volume of a rectangular room with dimensions 15 feet by 12 feet and walls with a height of 10 feet?

6. A farmhouse with 1,400 square feet of living space has 10-foot-high walls. What is the volume of the farmhouse?

7. A grocery store is 1,000 feet long, 800 feet wide, and has walls 12 feet high. What is the volume of the store?

8. A store in a mall is a cube with volume 8,000 cubic feet. What is the height of the ceiling?

Application: Packing and Moving

Use these volume formulas to solve the following problems.

Cube	$V = s^3$	where s is the side of the cube
Rectangular box	$V = \ell \times w \times h$	where ℓ, w, and h are the three dimensions of the box
Cylinder	$V = Bh$ $V = \pi r^2 h$	where B is the area of each round base, h is the height, $\pi \approx 3.14$, and r is the radius of the base
Cone	$V = \frac{1}{3}\pi r^2 h$	where $\pi \approx 3.14$, r is the radius of the base, and h is the distance between the tip of the cone and the base

1. Jones Moving Company was sent to the Smith residence to pack and move their furniture. The Smith children have a large supply of building blocks. The movers counted 156 blocks. Each cube block is 3 inches high. What is the volume of each building block? What is the total volume of all 156 blocks?

2. The movers want to pack the blocks in a rectangular box 3 feet long, 2 feet wide, and 1 foot high. What is the volume of the packing box in cubic feet?

3. Will the blocks all fit into one packing box? (Hint: Write both measurements in cubic inches.)

4. The movers have to carefully pack all of the Smith's drinking glasses. Each glass is 5 inches tall with a diameter of 3 inches. What is the volume of each drinking glass?

5. The Smith's endtables are shaped like a drum. Each table is 2 feet high and has a radius of 18 inches. What is the volume of each end table?

6. The Smith's refrigerator is $5\frac{1}{2}$ feet high, 3 feet wide, and 3 feet long. Find the volume of the refrigerator.

7. The Smith's kitchen table has cone-shaped legs. The movers removed the four legs and wondered if all four would fit into a long rectangular packing box 2 feet by 2 feet by 4 feet. Each table leg is 3 feet long with a radius of 2 inches. Find the volume of each table leg and the volume of the packing box.

8. The Smiths need to get their house ready to sell. They ordered a pile of gravel for their driveway. When it was delivered, it formed a cone that was 5 feet high with a diameter of 6 feet. What is the volume of the gravel?

Section 4 Cumulative Review

Rewrite each measurement so it uses the given unit.

1. 1 qt = _____ c

2. 2 pt = _____ qt

3. 1 L = _____ mL

4. _____ tsp = 1 tbsp

5. _____ fl oz = 1 tbsp

6. 1 c = _____ fl oz

7. 1 L _____ kL

8. 1 qt = _____ fl oz

9. _____ qt = 1 gal

Mark the measurement on each container.

10. 75 mL

11. $1\frac{1}{2}$ fl oz

12. $1\frac{3}{4}$ c

Circle the smaller measurement.

13. 1 L or 1 qt

14. 1 mL or 1 fl oz

15. 1 c or 1 qt

16. 1 L or 1 gal

17. 1 tsp or 1 c

18. 1 pt or 1 L

Underline the best estimate for the capacity of each object.

19. a dose of medicine   2 fl oz, 2 c, 2 pt

20. a serving of ice cream   8 mL, 8 c, 8 fl oz

21. a serving of coffee   1 qt, 1 mL, 1 c

22. oil for a lawn mower   1 gal, 1 qt, 1 tbsp

23. water for a punch recipe   1 kL, 1 L, 1 mL

Rewrite each measurement so it uses the given unit.

24. 6 c = _____ pt

25. _____ qt = 42 c

26. 4 gal = _____ qt

27. 6 fl oz = _____ tbsp

28. 750 mL = _____ L

29. 2.4 kL = _____ L

30. 18 qt = _____ gal

31. 28 qt = _____ c

32. _____ mL = 1.6 L

Write the abbreviation for each unit.

33. fluid ounce   _____

34. liter   _____

35. cup   _____

36. gallon   _____

37. milliliter   _____

38. quart   _____

Rewrite each measurement so it uses the given unit.

39. ____ mL = 2.4 L

40. 3 pt = ____ c

41. $2\frac{1}{2}$ gal = ____ qt

42. ____ fl oz = 3 tbsp

43. 6 gal = ____ pt

44. 18 c = ____ pt

45. 9 tsp = ____ tbsp

46. 15 c = ____ qt

Write the measurements from smallest to largest.

47. 3 c, 3 tsp, 3 pt   _____

48. 750 mL, 7 L, 1500 mL   _____

49. $6\frac{1}{2}$ qt, 14 qt, 28 c   _____

Underline the largest amount.

50. 6 c, 2 qt, 3 pt   51. 2 L, 6,500 mL, 1 kL   52. 1 fl oz, 10 tsp, 4 tbsp

Circle the best estimate for each amount.

53. amount of blood to test cholesterol   50 L   700 mL   250 mL

54. amount of car oil for a sports car   5 gal   5 qt   5 c

55. amount of water to take a shower   15 qt   15 kL   15 gal

Perform each operation.

56. 6 qt 2 c
    − 3 qt 3 c

57. $\dfrac{\text{6 gal 2 qt}}{4}$

58. 7 L 600 mL
    + 3,450 mL

59. 3 pt 1 c
    ×   5

Solve each problem.

60. Find the volume of a cube with 6-inch sides.

61. Find the volume of a rectangular solid with dimensions $h = 6$ inches, $w = 7$ inches, and $\ell = 8$ inches.

62. What is the volume of a cylinder that is 10 meters high and has a diameter of 6 meters?

63. Henry wants to cover an area 20 feet long and 8 feet wide with a layer of concrete 6 inches thick. What is the volume of the concrete?

64. Mark had to add 7 cups of oil to his car. How many quarts is that?

65. A painter said she needed $1\frac{1}{2}$ gallons of paint to finish her job. How many quarts is that?

66. What is the volume of a rectangular container that is 4 centimeters by 5 centimeters by 6 centimeters?

67. A recipe calls for a $1\frac{1}{2}$ pints of whole tomatoes. How many cups is that?

68. Josie is supposed to take 6 fluid ounces of medicine a day. How many tablespoons is that?

69. The volume of a box is 1,400 cubic inches. The width is 7 inches and the length is 10 inches. What is the height of the box?

70. What is the volume of a cone that is 6 inches high and has a radius of 3 inches?

71. Jason works as a lab technician. He had to get 250 milliliters of blood from a patient for a test. How many liters of blood is that?

List the names of the units.

72. List the customary units and the metric units for measuring weight.

_____

_____

73. List the customary units and the metric units for measuring capacity.

_____

_____

74. List the customary units and the metric units for measuring length.

_____

_____

Write the measurement shown on each ruler.

75.

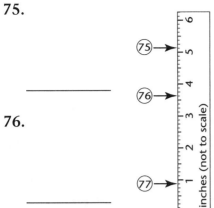

76. _____

77.

_____

78.

79. _____

80. _____

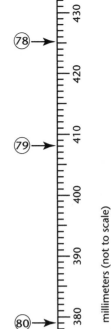

Find each indicated length.

81.

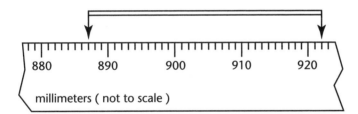

_____

82.

_____

Find the complement of each angle.

83.  _____

84.  _____

Find the supplement of each angle.

85.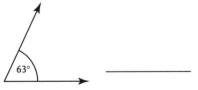

_____

86.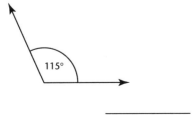

_____

Messenger and Courier Services

In cities, you see them in cars, on bicycles, and even on in-line skates. They are messengers and they are delivering small packages, documents, or other information from one company to another. On bicycles or skates, they wear a helmet, gloves, and a uniform, and they carry a dispatch case containing their packages, paperwork, and a two-way radio. Messengers also maintain records of deliveries and obtain signatures from the persons who receive items.

The radio is the messengers' communication link with their dispatcher. The dispatcher has to keep track of each messenger and package. Also, the dispatcher has to know the name of each client who has a package for pick-up and the name of each person who signs for and receives it. The dispatcher has to record times and durations: the time of each client's call, the time elapsed between the client's call and the package pick-up, and the time elapsed between pick-up and delivery.

Clients need their packages delivered as quickly as possible. Messengers and couriers are paid according to the number of deliveries they make, so they always move quickly. If you need directions, ask a messenger. If he or she can take the time to help you, you will learn the quickest way to get to your destination. Messengers maximize the number of their deliveries by finding the fastest route between any two places in their city. So, they need a good knowledge of the geographic area they serve and a good sense of direction.

Suppose you are a dispatcher for a messenger service. Here are some of the questions you ask yourself each day. For each question, think about how you would answer it.

- How many messengers are needed today? In what parts of the city are they needed the most?

- What is the distance between the pick-up and delivery point for a package?

- How long will it take a messenger to travel that distance?

- How many miles per hour must the messenger drive?

- If two clients who have packages for pick-up are near each other, should one messenger pick up both packages? If so, which package should be picked up first? Which package should be delivered first?

> Many of the questions addressed by workers in messenger and package-delivery services involve speed or velocity and time. In this lesson, you will learn about units for measuring time and units for measuring velocity. Also, you will investigate and solve problems related to time and velocity.

Career Corner

There are many companies that deliver packages, including the U.S. Postal Service. Here are some related jobs in the field of messenger and courier services.

| route drivers | receiving clerks | traffic clerks |
| postal clerks | shipping clerks | mail carriers |

Can you think of any others?

Units of Time

When you measure a brief amount of time, you might use **seconds** or **minutes** as the unit. Other familiar units of time are **hours**, **days**, **weeks**, and **months**. Longer units of time are **years**, **decades**, and **centuries**. Here are some of the relationships among these units of time.

1 minute (min) = 60 seconds (sec)	1 yr = 52 weeks
1 hour (hr) = 60 min	1 decade = 10 years
1 day = 24 hr	1 century = 100 years
1 week = 7 days	
1 year (yr) = 12 months (mo)	

Here are some examples of approximate time.

Saying the words "ten thousand" takes about 1 second.

Washing your hands takes about 1 minute.

Baking a loaf of bread takes about 1 hour.

Name three things that are measured in the given unit.

1. seconds

_____

2. weeks

_____

3. hours

_____

Write the unit that is most appropriate to measure each amount of time.

4. length of time to pay off a mortgage _____

5. time needed to wallpaper a kitchen _____

6. time needed to cook a hamburger _____

7. how long it takes to fill a car's gas tank _____

8. how long someone studies to become a doctor _____

9. time on a train between Seattle and San Francisco _____

10. amount of time to eat three pizza slices _____

11. length of time in one school semester _____

Circle the time that is more appropriate for each event.

12. time to fly 1,200 miles
2 days or 2 hours

13. time to work the early shift
8 minutes or 8 hours

14. time to wash dinner dishes
20 minutes or 20 seconds

15. time to grow tomatoes
4 days or 4 weeks

16. time it takes to build a store
15 days or 15 months

17. time it takes to jog 2 miles
20 minutes or 20 hours

18. time it takes to lose 10 pounds
12 weeks or 12 hours

19. time it takes to bake a cake
40 minutes or 40 hours

A stopwatch can be used to measure short periods of time, such as how long a runner takes to complete a race. A period of time is often called an **interval** or a **duration**.

Example

Janice used her stopwatch to time a runner in a 400-meter race. She pressed the button once to start the watch, and then again to stop the watch. According to her stopwatch, how long did it take the runner to complete the 400-meter race?

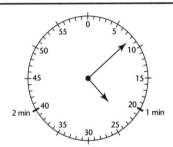

Step 1. Decide what units are used.

minutes and seconds

Step 2. See where the arrow has stopped.

The watch shows 1 minute and 8 seconds.

Answer: The runner finished the 400-meter race in 1 minute and 8 seconds.

Write the duration shown on each stop watch.

20.

duration: _____

21.

duration: _____

22.

duration: _____

Write the duration shown on each stop watch.

23.

duration: _____

25.

duration: _____

27.

duration: _____

24.

duration: _____

26.

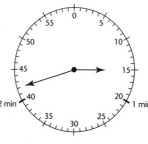

duration: _____

28.

duration: _____

Mark the duration given on the face of the stop watch.

29.

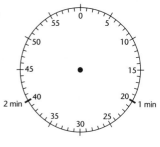

1 min 10 sec

31.

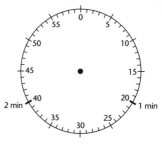

8 seconds

33.

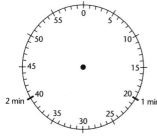

2 min 37 sec

30.

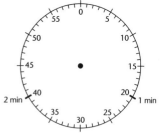

2 min 15 sec

32.

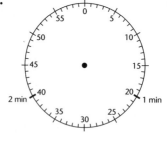

1 min 51 sec

34.

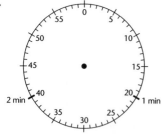

0 min 24 sec

Estimating Time

Estimate how long it would take you to do each activity.

1. walk to work or school

   _____

2. iron a shirt

   _____

3. scrub the kitchen floor

   _____

4. watch a soccer game

   _____

5. take a shower

   _____

6. walk 4 miles

   _____

7. take the trash out

   _____

8. cook a 15-pound turkey

   _____

9. read a magazine

   _____

10. wash a car

   _____

Complete each measurement by writing an appropriate unit of time.

11. vacation after being with   3 _____
    a company for 10 years

12. lunch break   30 _____

13. good nights' rest   8 _____

14. time to get a drink at a water fountain   20 _____

15. time to buy groceries   40 _____

16. time for snow drifts to melt   6 _____

17. time for a movie $2\frac{1}{2}$ _____

18. time worked in one week   45 _____

19. time to earn a nurse's license   3 _____

20. time of summer break for students   10 _____

Changing Among Units of Time

When you use a smaller unit to measure a time interval, the number of units in your measurement will be larger. When you use a larger unit, the number of units in your measurement will be smaller.

Here is a review of the relationships among seconds, minutes, hours, and days.

1 min	=	60 sec
1 hr	=	60 min
1 day	=	24 hr

— Example 1

Connie's dog sleeps 12 hours a day. How many minutes does Connie's dog sleep each day?

Step 1. Write the problem.   12 hr = ___ min

Step 2. Write the relationship between minutes and hours.   1 hr = 60 min

Step 3. We want to go from 1 hour to 12 hours. So multiply both sides of the equation in Step 2 by 12.   12 hr = 720 min

Answer: Connie's dog sleeps 720 minutes every day.

— Example 2

How many minutes are there in 3 days?

Step 1. Write the problem.   3 days = ___ min

Step 2. Write the relationship between days and hours. Each hour is the same as 60 minutes, so multiply $24 \times 60 = 1{,}440$ to find the number of minutes in one day.
     1 day = 24 hr
     1 day = (24)(60) min
     1 day = 1,440 min

Step 3. We want to go from 1 day to 3 days, so multiply both sides of the equation "1 day = 1,440 min" by 3.   3 days = 4,320 min

Answer: There are 4,320 minutes in 3 days.

Rewrite each measurement so it uses the given unit.

1. 120 hr = _____ sec

2. 5 days = _____ sec

3. 72 days = _____ hr

4. 2 days = _____ min

5. 4 hr = _____ min

6. 10 min = _____ sec

7. 7 days = _____ min

8. 3 hr = _____ min

9. 45 min = _____ sec

Sometimes a time measurement contains a mixed number. Here is an example.

Example 3

Change $3\frac{1}{2}$ days to hours.

Step 1.	Write the problem.	$3\frac{1}{2}$ days = ___ hr
Step 2.	Write the relationship between days and hours.	1 day = 24 hr
Step 3.	We want to go from 1 day to $3\frac{1}{2}$ days. So multiply both sides of the equation in Step 2 by $3\frac{1}{2}$.	$3\frac{1}{2}$ days = $24 \times 3\frac{1}{2}$ $= (24)(3) + (24)(\frac{1}{2})$ $= 72 + 12$ $= 84$

Answer: $3\frac{1}{2}$ days = 84 hours

Rewrite each measurement so it uses the given unit.

10. 2.5 hr = ____ min   **13.** 4.2 hr = ____ min   **16.** $2\frac{3}{4}$ hr = ____ min

11. $6\frac{1}{4}$ days = ____ hr   **14.** 2.5 days = ____ hr   **17.** $1\frac{1}{2}$ days = ____ min

12. 12.3 min = ____ sec   **15.** 3.8 min = ____ sec   **18.** 5.6 min = ____ sec

Example 4

Leon is studying for the GED writing test. He will have 120 minutes to finish the test. How many hours will Leon have to work on the test?

Step 1.	Write the problem.	120 min = ____ hr
Step 2.	Write the relationship between minutes and hours.	60 min = 1 hr
Step 3.	We want to go from 60 minutes to 120 minutes. So multiply both sides of the equation in Step 2 by 2.	120 min = 2 hr

Answer: Leon will have 2 hours to work on the GED writing test.

Example 5

Change 180 seconds to minutes.

Step 1.	Write the problem.	180 sec = ___ min
Step 2.	Write the relationship between seconds and minutes.	60 sec = 1 min
Step 3.	We want to go from 60 seconds to 180 seconds. So multiply both sides of the equation in Step 2 by 3.	180 sec = 3 min

Answer: 180 seconds = 3 minutes

Rewrite each measurement so it uses the given unit.

19. 3,600 sec = ___ min   **22.** 168 hr = ___ days   **25.** 300 sec = ___ min

20. 48 hr = ___ days   **23.** 600 sec = ___ min   **26.** 240 min = ___ hr

21. 120 min = ___ hr   **24.** 96 hr = ___ days   **27.** 1,440 min = ___ hr

When you change from a smaller unit to a larger unit, the new measurements may contain a mixed number or a decimal. Here are two examples.

___ Example 6 _____

Change 105 minutes to hours.

Step 1. Write the problem.   105 min = ___ hr

Step 2. Write the relationship   60 min = 1 hr
between minutes and hours.

Step 3. Set up a proportion with minutes $\dfrac{105}{?} = \dfrac{60}{1}$
on the top and hours on the bottom.
Use "?" to represent the unknown
number of hours.

Step 4. Find the cross-products and $? \times 60 = 105 \times 1$
solve for the unknown number. $? = \dfrac{105}{60} = \dfrac{7}{4} = 1\dfrac{3}{4}$

Answer: 105 minutes = $1\dfrac{3}{4}$ hours

___ Example 7 _____

Change 204 hours to days.

Step 1. Write the problem.   204 hr = ___ days

Step 2. Write the relationship between   24 hr = 1 day
hours and days.

Step 3. Set up a proportion with hours $\dfrac{204}{?} = \dfrac{24}{1}$
on the top and days on the bottom.
Use "?" to represent the unknown
number of days.

Step 4. Find the cross-products and $? \times 24 = 204 \times 1$
solve for the unknown number. $? = \dfrac{204}{24} = 8.5$

Answer: There are 8.5 days in 204 hours.

Change each measurement to a mixed number.

28. 90 sec = ___ min

29. 2,190 hr = ___ days

30. 75 min = ___ hr

31. 36 hr = ___ days

32. 105 min = ___ hr

33. 80 sec = ___ min

34. 150 sec = ___ min

35. 50 hr = ___ days

36. 100 min = ___ hr

Solve each problem.

37. Jerry has to be in school for 8 hours each week day. How many minutes does he spend in school each week?

38. Tom bought a car by borrowing the money from his bank. He received a 36-month loan. How many years will it take to pay off the bank loan?

39. Jim works 8 hours a day, 5 days a week. How many hours does he work in one week?

40. Jim is planning on taking a two-week vacation. How many hours of work will he miss while on his vacation?

41. Sally packages carpentry nails at a factory. She can package 100 nails in a box every 15 seconds. How many minutes will it take her to package 200 nails?

42. A job fair was held on May 9th. The fair was open to the public for 480 minutes. For how many hours was the fair open?

43. Michelle spends 3 hours per week in an English 101 class. How many minutes is she in class each week?

44. Michelle's English class includes a one-hour lab twice a week. How many minutes is she in lab each week?

45. Joel connected his computer to an Internet service. His computer ran for 432 hours the first month. For how many days was he connected to the Internet?

46. Taneisha found a recipe for banana bread. Cooking time is 90 minutes. How much time should the bread bake? (Your answer should be in hours and minutes.)

47. Jean watched a semi-final basketball game on television. She saw that the teams took four time outs, each one 45 seconds long. How much total time did they take in time outs?

48. Cheri needed to take her child's temperature. Her thermometer beeped after 2 minutes to let her know it had a reading. How many seconds did it take to find out if her child had a fever?

Adding and Subtracting Measurements of Time

When you add or subtract time measurements, you may have to carry or borrow between the different units. Here are two examples.

Example 1

Barb worked 6 hours and 50 minutes on Tuesday and 5 hours 30 minutes on Thursday. How much total time did she work on the two days?

Step 1. Line up the measurements, putting like units under like units.

$$\begin{array}{r} 6 \text{ hr } 50 \text{ min} \\ + \ 5 \text{ hr } 30 \text{ min} \\ \hline \end{array}$$

Step 2. Add the minutes and add the hours.

$$\begin{array}{r} 6 \text{ hr } 50 \text{ min} \\ + \ 5 \text{ hr } 30 \text{ min} \\ \hline 11 \text{ hr } 80 \text{ min} \end{array}$$

Step 3. 80 minutes is 1 hour and 20 minutes. Rewrite the sum.

$$11 \text{ hr} + 80 \text{ min} = 11 \text{ hr} + 1 \text{ hr} + 20 \text{ min}$$
$$= 12 \text{ hr} + 20 \text{ min}$$

Answer: Barb worked 12 hours and 20 minutes on the two days.

Add.

1. $\begin{array}{r} 2 \text{ years } \ 7 \text{ months} \\ + \ 8 \text{ years } 10 \text{ months} \\ \hline \end{array}$

2. $\begin{array}{r} 8 \text{ weeks } 4 \text{ days} \\ + \ 10 \text{ weeks } 6 \text{ days} \\ \hline \end{array}$

3. $\begin{array}{r} 10 \text{ min } 43 \text{ sec} \\ + \ 12 \text{ min } 38 \text{ sec} \\ \hline \end{array}$

Example 2

Subtract 6 minutes 45 seconds from 12 minutes 20 seconds.

Step 1. Line up the measurements, putting like units under like units.

$$\begin{array}{r} 12 \text{ min } 20 \text{ sec} \\ - \ \ 6 \text{ min } 45 \text{ sec} \\ \hline \end{array}$$

Step 2. One minute is the same as 60 seconds. Rewrite the top number as 11 minutes + 60 seconds + 20 seconds, or 11 minutes 80 seconds.

$$\begin{array}{r} 11 \text{ min } 80 \text{ sec} \\ - \ \ 6 \text{ min } 45 \text{ sec} \\ \hline \end{array}$$

Step 3. Subtract the seconds and subtract the minutes.

$$\begin{array}{r} 11 \text{ min } 80 \text{ sec} \\ - \ \ 6 \text{ min } 45 \text{ sec} \\ \hline 5 \text{ min } 35 \text{ sec} \end{array}$$

Answer: The result of the subtraction is 5 minutes 35 seconds.

Subtract.

4. $\begin{array}{r} 6 \text{ hours } 40 \text{ min} \\ - \ 3 \text{ hours } 15 \text{ min} \\ \hline \end{array}$

5. $\begin{array}{r} 7 \text{ days } \ 6 \text{ hours} \\ - \ 2 \text{ days } 12 \text{ hours} \\ \hline \end{array}$

6. $\begin{array}{r} 12 \text{ years } 18 \text{ weeks} \\ - \ 7 \text{ years } 10 \text{ weeks} \\ \hline \end{array}$

Add or subtract.

7. 17 hr   0 min
    − 6 hr 45 min

10.   6 hr
    + 3 hr 25 min

13.   28 min 30 sec
    − 16 min 48 sec

8.   32 hr 7 min
    − 18 hr

11.   16 days 10 hr
    + 12 days   8 hr

14.   3 mo
    − 2 mo 1 wk

9.   6 days   7 hr
    − 2 days 20 hr

12.   16 min 48 sec
    + 32 min 45 sec

15.   21 days 17 hr
    + 6 days 18 hr

Solve each problem.

16. Sol earned 4 hours 30 minutes vacation last week and 6 hours 45 minutes vacation this week. How much vacation time did he earn in the 2-week period?

17. During the past 18 years, Jay has worked at an auto parts store. Because of a long illness, he did not work 1 year 3 months during that 18-year period. How many years has he worked there?

18. A computer keeps very accurate records of how long a telemarketer is on the phone. Ronette's records show calls of 3 minutes 28 seconds, 2 minutes 47 seconds, and 5 minutes 55 seconds. For how long was she on the phone?

19. Jackie was given a 6-hour telephone card as a gift, which allows her to charge long distance calls. She has used 2 hours 39 minutes. How much time does she have left on the card?

20. Kurt worked these hours during the week: 7 hours 48 minutes; 8 hours 21 minutes; 6 hours 57 minutes; 8 hours 46 minutes; and 8 hours 38 minutes. How many total hours did he work during the week?

Multiplying and Dividing Measurements of Time

For some problems you can multiply or divide a time measurement by a number. For example, if you know the total time someone worked during a week, you can divide to find out the average time worked each day. If your commute to school or work takes the same number of minutes each day, you can multiply to find out how much time you spend commuting each month.

Example 1

Elena worked at the grocery store 5 days last week. Her total time worked last week was 42 hours. What was the average length of her workday, in hours and minutes?

Step 1. Decide which math operation to use.

The operation is division.

Step 2. Write the problem and divide.

$\frac{42}{5} = 8\frac{2}{5}$ hours

Step 3. To change the fraction $\frac{2}{5}$ to minutes, multiply $\frac{2}{5}$ times the number of minutes in an hour.

$\frac{2}{5} \times 60 = 24$

Answer: Elena's average workday last week was 8 hours 24 minutes.

Divide.

1. 6 hours 10 minutes ÷ 5

2. 13 weeks 5 days ÷ 6

3. 8 months 20 days ÷ 4
   (Use 1 month = 30 days.)

4. 46 hours 20 minutes ÷ 5

Example 2

Multiply 6 weeks 3 days by 5.

Step 1. Write the problem.

$$\begin{array}{r} 6 \text{ weeks } 3 \text{ days} \\ \times \qquad\qquad 5 \\ \hline \end{array}$$

Step 2. Multiply the number of days and the number of weeks by 5.

$$\begin{array}{r} 6 \text{ weeks } \quad 3 \text{ days} \\ \times \qquad\qquad 5 \\ \hline 30 \text{ weeks } 15 \text{ days} \end{array}$$

Step 3. 15 days is the same as 2 weeks and 1 day.

30 weeks plus 15 days
= 30 weeks plus 2 weeks plus 1 day
= 32 weeks 1 day

Answer: The product is 32 weeks 1 day.

Multiply or divide.

5. 2 years 4 months × 4

6. 15 min 20 sec × 5

7. 3 years 8 month ÷ 2

8. 3 days 8 hours ÷ 4

9. 16 years 4 months × 3

10. 6 months 3 days × 3

11. 6 days 14 hours × 5

12. 12 days 18 hours ÷ 3

13. 6 min 38 sec × 5

14. 7 hours 12 minutes × 4

Solve each problem.

15. Thomas works in a copy center. He figures an average job takes him 5 minutes 20 seconds to complete. If he did 22 jobs on Tuesday, how many hours, minutes, and seconds did he work?

16. If Chu averages 8 hours and 20 minutes daily on his job, how many total hours does he work in a five-day work week? How many total hours does he work in a six-day work week?

17. Kym is a nurse's aid in a nursing home. If she works with 8 residents during the 48 minutes 24 seconds that she was giving out medications, how much time did she average with each resident?

18. During the past 10 years 3 months, Ryan has had three different jobs. What is his average time for each job?

19. Emil decided he needed to exercise daily. If he does aerobic exercises for 45 minutes each day, including weekends, how many hours does he exercise during a five-week period?

20. The water pump at the Jetson's house runs about $2\frac{1}{2}$ hours daily. How many hours does the pump run weekly?

Comparing and Ordering Time Measurements

In order to compare the measurements of two time intervals, it is easier if the measurements use the same unit.

— Example 1

The directions on a box of cake mix say to beat the batter for 2 minutes. Joel watched the second hand on the kitchen clock and stopped mixing after 90 seconds. Did he beat the cake mix batter long enough?

Step 1.	Write the two measurements.	90 sec, 2 min
Step 2.	Rewrite one of the measurements so they have the same unit. This example uses seconds.	90 sec, 120 sec
Step 3.	Compare the measurements.	90 sec < 120 sec
Answer:	Since 90 seconds is less than 2 minutes, Joel did not beat the cake mix batter long enough.	

— Example 2

Which is shorter, 3 weeks or 18 days?

Step 1.	Write the two measurements.	3 weeks, 18 days
Step 2.	Rewrite one of the measurements so they have the same unit. This example uses days.	21 days, 18 days
Step 3.	Compare the measurements.	21 days > 18 days
Answer:	The shorter period of time is 18 days.	

Circle the longer period of time.

1. 3 days or 80 hours

2. 360 minutes or 4 hours

3. 10 weeks or 3 months

4. 2 years or 22 months

5. 3 hours or 30,000 seconds

6. 72 hours or $3\frac{1}{2}$ days

7. 8 months or 274 days

8. 6 weeks or 50 days

9. 3 years or 160 weeks

10. 248 minutes or 4 hours

11. 6 weeks or 2 months

12. $3\frac{1}{2}$ years or 65 weeks

Fill in each blank with <, >, or =.

13. 2 days ____ 48 hours

14. 3 weeks ____ 24 days

15. 6 years ____ 56 months

16. 18 hours ____ 475 minutes

17. 70 months ____ 26 weeks

18. $6\frac{1}{2}$ hours ____ 420 minutes

19. 200 minutes ____ $3\frac{1}{2}$ hours

20. 17 weeks ____ 4 months

21. $3\frac{1}{2}$ years ____ 42 months

22. 180 seconds ____ 5 minutes

23. 30 hours ____ 2 days

24. 4 days ____ 96 hours

Write the measurements from shortest to longest.

25. 3 minutes, 200 seconds, 45 seconds

_____

26. $2\frac{1}{2}$ days, 65 hours, 420 minutes

_____

27. 36 weeks, 10 months, 1 year

_____

28. $2\frac{1}{2}$ years, 40 months, 85 weeks

_____

Solve each problem.

29. Mary has 13 vacation days. Jose has 3 weeks. Who has more vacation days?

30. The local furniture store was advertising a 36-hour marathon sale. Was the sale more or less than 2 days?

31. Manuel volunteered as an aide at the nursing home 40 hours last month. Is his volunteer time more or less than 3,000 minutes?

32. The supervisor of a glass-cutting department thought a job should take 15 minutes. Sam figures he can do the job in a quarter hour. Is his time more than, less than, or the same as what the supervisor says is adequate?

33. At an adult education class, Jack spends 6 hours a week working on computer skills. Tran thinks he spends about 400 minutes a week on his home computer. Who spends more time on a computer?

Focus on Algebra: Velocity

When an object or person is moving, we can measure its **velocity**. Velocity is the same as speed or rate. When an object is traveling at a constant rate or velocity, we can use a formula to calculate that velocity. If an object travels a distance d in a time t, then the average velocity r can be calculated from the formula $r = \dfrac{d}{t}$.

—— Example 1 ——

On vacation with her family, Maria drove 180 miles in 3 hours. What was the average velocity for the trip?

Step 1.	Write the formula.	$r = \dfrac{d}{t}$
Step 2.	Substitute numbers from the problem into the formula.	$r = \dfrac{180}{3}$
Step 3.	Divide to simplify the fraction.	$r = 60$
Answer:	The rate is 60 miles per hour (mph).	

Use the formula $r = \dfrac{d}{t}$ to find each missing velocity.

1. $r =$ _____
   $d = 7.5$ mi
   $t = 3$ hr

2. $r =$ _____
   $d = 1{,}800$ mi
   $t = 6$ hr

3. $r =$ _____
   $d = 180$ mi
   $t = 1.5$ hr

You can use the formula $r = \dfrac{d}{t}$ to find the distance if you know the rate and time.

—— Example 2 ——

On a state highway, the speed limit is 55 mph. If Jon drove for 3 hours at the speed limit, how far did he drive?

Step 1.	Write the distance formula.	$r = \dfrac{d}{t}$
Step 2.	Substitute numbers from the problem into the formula. Be sure to substitute each number for the correct **variable**, or letter.	$55 = \dfrac{d}{3}$
Step 3.	Multiply both sides of the equation in Step 2 by 3.	$d = 165$
Answer:	Jon drove 165 miles in three hours.	

Use the formula $r = \dfrac{d}{t}$ to find each missing distance.

4. $r = 400$ mph
   $d =$ _____
   $t = 22.5$ hr

5. $r = 10$ mph
   $d =$ _____
   $t = 6\frac{1}{2}$ hr

6. $r = 60$ mph
   $d =$ _____
   $t = 2\frac{3}{4}$ hr

There are times when you might want to know how long a trip will take at a particular average velocity. For example, a truck driver or moving van company might need to know how long a 414 mile trip would take at an average velocity of 52 miles per hour.

Example 3

Jarod is about to drive his truck 416 miles. From experience, he knows his average velocity will be about 52 mph. How long will the trip take?

Step 1. Write the formula.

$$r = \frac{d}{t}$$

Step 2. Substitute numbers from the problem into the formula.

$$52 = \frac{416}{t}$$

Step 3. Solve the equation.

$$52t = 416$$
$$t = \frac{416}{52} = 8$$

Answer: The trip will take about 8 hours.

Use the formula $r = \dfrac{d}{t}$ for each exercise.

7. $r = 61$ mph
   $d = 366$ mi
   $t = $ _____

8. $r = $ _____
   $d = 1,050$ mi
   $t = 3\frac{1}{2}$ hr

9. $r = 60$ mph
   $d = $ _____
   $t = 4\frac{1}{2}$ hr

10. $r = 150$ mph
    $d = 300$ mi
    $t = $ _____

11. $r = 60$ mph
    $d = 135$ mi
    $t = $ _____

12. $r = $ _____
    $d = 480$ mi
    $t = 1\frac{1}{3}$ hr

13. $r = 40$ mph
    $d = 220$ mi
    $t = $ _____

14. $r = 65$ mph
    $d = $ _____
    $t = 8$ hr

15. $r = 50$ mph
    $d = 600$ miles
    $t = $ _____

Solve each problem.

16. A 747 jet made a 1,300-mile flight in $3\frac{1}{4}$ hours. What was the average velocity for the jet?

17. Jason rode his bike $27\frac{1}{2}$ miles in 2 hours. What was his average velocity?

18. Sam drove a van 243 miles at an average velocity of 54 mph. How many hours did he drive?

19. Jackie made a 150-mile trip in $2\frac{1}{2}$ hours. What was her average speed?

20. How many hours will it take Fred to drive 168 miles at an average velocity of 48 mph?

Application: Time Zones

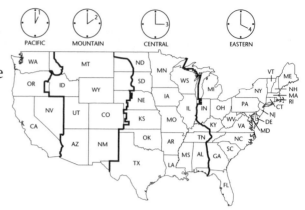

Have you ever called a relative on the East coast or West coast and been asked "Why are you calling at this hour? You woke me up." You look at your clock and it's only 9:00 P.M.

The different regions of the country have different **time zones**. The different time zones mean that the clocks in our homes, schools, and businesses are coordinated with the apparent position of the sun.

Most of the United States is covered by four regions: Pacific, Mountain, Central, and Eastern. On the map above, you can see the extent of each region. As you change regions from west to east, the time on the clock is moved ahead by one hour. For example, at 9 A.M. in the Pacific region, it is 10 A.M. in the Mountain region, 11 A.M. in the Central region, and noon in the Eastern region.

Use the time zone map to determine each of the following times.

1. 8 A.M. Pacific = _____ Central

2. Noon Mountain = _____ Pacific

3. 3 P.M. Eastern = _____ Mountain

4. 5:30 P.M. Central = _____ Eastern

5. 7 P.M. Pacific = _____ Eastern

6. 6:45 A.M. Eastern = _____ Pacific

7. 1:15 P.M. Mountain = _____ Pacific

8. 11 A.M. Mountain = _____ Pacific

9. 10:40 P.M. Central = _____ Mountain

10. Midnight Eastern = _____ Mountain

Solve each problem.

11. George lives in New York. He needs to talk to his business partner at 7:30 A.M., California time. What time should he call from New York?

12. Maria lives in Texas. At 9:00 P.M. she called her mother in Utah. What time was it at her mother's home?

13. Julia flew from Albuquerque, NM, to San Francisco, CA, an hour's flight. She left Albuquerque at 10:15 A.M. What time did she arrive in San Francisco?

14. Sam lives in Florida. He called his aunt in Seattle, which is on Pacific time. If he called at 5 P.M. Florida time, what time was it in Seattle?

15. Antoine needs to fly to Atlanta, GA, from Chicago, IL. If he leaves Chicago at 6 A.M., what time is that in Atlanta?

16. Jimmy called to California from Boston, Massachusetts. How many hours difference is there in the time between the two time zones?

Application: Wind Velocity

The velocity or speed of wind is measured by an anemometer. In 1805, Sir Francis Beaufort developed a wind scale to measure the effect wind had on a ship's sails. Wind is measured 10 meters (about 40 feet) above the ground.

			Beaufort Wind Scale
Beaufort Number	**Name**	**Miles per hour**	**Effect on land**
0	Calm	less than 1	Calm; smoke rises vertically.
1	Light Air	1 - 3	Weather vanes inactive; smoke drifts with air.
2	Light Breeze	4 - 7	Weather vanes active; wind felt on face; leaves rustle
3	Gentle Breeze	8 - 12	Leaves and small twigs move; light flags extend.
4	Moderate Breeze	13 - 18	Small branches sway; dust and loose paper blow about.
5	Fresh Breeze	19 - 24	Small trees sway; waves break on inland waters.
6	Strong Breeze	25 - 31	Large branches sway; umbrellas difficult to use.
7	Moderate Gale	32 - 38	Whole trees sway; difficult to walk against wind.
8	Fresh Gale	39 - 46	Twigs break off trees; walking against wind very difficult.
9	Strong Gale	47 - 54	Slight damage to buildings; shingles blown off roof.
10	Whole Gale	55 - 63	Trees uprooted; considerable damage to buildings.
11	Storm	64 - 73	Widespread damage; very rare occurrence.
12 - 17	Hurricane	74 and above	Violent destruction.

Find the Beaufort number for each exercise.

1. wind at 21 mph

2. light air

3. wind at 70 mph

_____   _____   _____

Find the range of miles per hour for each exercise.

4. a strong breeze

5. Beaufort number 7

6. a gentle breeze

_____   _____   _____

Describe the effect on land for each wind velocity.

7. wind of 49 mph

10. Beaufort number 10

_____   _____

8. fresh breeze

11. wind of 72 mph

_____   _____

9. calm air

12. Beaufort number 4

_____   _____

Application: Miles Per Gallon

How far can your car go on a gallon of gasoline? You can calculate the **miles per gallon** (mpg) if you know the number of miles you drove and the number of gallons of gasoline you used.

— Example 1 —————————————————————————

Rollie put 11.6 gallons of gas in his car. He knew he had driven 290 miles since his last fill-up. How many miles per gallon (mpg) did he average?

Step 1. Set up a proportion using the ratio $\dfrac{\text{miles}}{\text{gallons}}$.

$$\frac{290}{11.6} = \frac{?}{1}$$

Step 2. Write the cross-products for the proportion.

$$? \times 11.6 = 290 \times 1$$

Step 3. Divide each side by 11.6.

$$? = \frac{290}{11.6} = 25$$

Answer: Rollie got 25 miles per gallon on the trip.

Find the number of miles per gallon for each exercise.

1. 280 miles driven
   14 gallons of gas
   mpg = _____

2. 486 miles driven
   16 gallons of gas
   mpg = _____

3. 114 miles driven
   6 gallons of gas
   mpg = _____

4. 142.8 miles driven
   6.8 gallons of gas
   mpg = _____

5. 375 miles driven
   12.5 gallons of gas
   mpg = _____

6. 1,700 miles driven
   70 gallons of gas
   mpg = _____

If you know how many miles per gallon a car gets, you can relate the number of miles you can drive to the number of gallons of gasoline you will need.

— Example 2 —————————————————————————

Juan's car usually gets about 22 mpg. If he has 7 gallons of gas in his car, how far can he expect to drive before he needs more gasoline?

Step 1. Set up a proportion using the ratio $\dfrac{\text{miles per gallon}}{\text{gallons}}$.

$$\frac{22}{1} = \frac{?}{7}$$

Step 2. Write the cross-products.

$$? \times 1 = 22 \times 7$$

Step 3. Multiply.

$$? = 154$$

Answer: Juan can expect to drive 154 miles before he needs more gasoline.

Find the distance that can be driven on the given amount of gas and the known miles per gallon. (If necessary, round your answer to the nearest tenth.)

7. has 8 gal of gas
   24 mpg
   distance = _____

9. has 14.5 gal of gas
   18.4 mpg
   distance = _____

11. has 40 gal of gas
    19.5 mpg
    distance = _____

8. has 6.7 gal of gas
   31 mpg
   distance = _____

10. has 6.2 gal of gas
    25.8 mpg
    distance = _____

12. has 27 gal of gas
    30 mpg
    distance = _____

Solve each problem.

13. Chu's car averages 21 mpg. He has 8.6 gallons of gas in his car. How far can he expect to drive before he needs more gasoline?

14. Dina's truck started a trip with a full tank of gas. She drove 1,800 miles, and then needed 90 gallons to refill her gas tank. What is the mpg for the truck?

15. For one part of a trip, Sara started with an odometer reading of 24,586 miles. On her first stop to fill the tank with gas, the odometer showed 24,886 miles. She needed 10 gallons of gas to fill the tank. Figure out the mpg for this part of her trip. (Hint: Subtract the odometer readings to get the mileage.)

16. Rod's car gets 27 mpg. If he used 46 gallons of gas during a three-day trip, how many miles did he travel?

17. Yuko put 14 gallons of gas in her car on Monday. If she usually gets 31 mpg, how far can she drive before she needs more gas?

18. Ryan's mother lives 250 miles from his home. He has 12 gallons of gas in his car. The car usually gets 26 mpg. Does he have enough gas to drive to his mother's home and return?

19. Samuel delivers supplies to medical centers. The longest drive he has is 220 miles each way. If his delivery van holds 18 gallons of gas and averages 23 mpg, will he have to buy gas on the trip?

20. Jackie's sports car averages 14 mpg. How far can she expect to drive on 12 gallons of gas?

Application: Reading a Bus Schedule

Use this bus schedule to answer the following questions.

Eastbound							Westbound						
Crossroads	99-Maple	90-Maple	61-Maple	40-Cuming	23-Burt	16-Jackson	16-Jackson	23-Burt	40-Cuming	61-Maple	90-Maple	99-Maple	Crossroads
7:16	---	7:26	7:35	7:45	7:50	7:58	6:32	6:40	6:45	6:55	7:04	---	7:14
8:01	---	8:11	8:20	8:30	8:35	8:43	7:17	7:25	7:30	7:40	7:49	---	7:59
8:46	---	8:56	9:05	9:15	9:20	9:28	8:02	8:10	8:15	8:25	8:34	---	8:44
9:31	---	9:41	9:50	10:00	10:05	10:13	8:47	8:55	9:00	9:10	9:19	---	9:29
10:16	---	10:26	10:35	10:45	10:50	10:58	9:32	9:40	9:45	9:55	10:04	---	10:14
11:01	---	11:11	11:20	11:30	11:35	11:43	10:17	10:25	10:30	10:40	10:49	---	10:59
11:46	---	11:56	12:05	12:15	12:20	12:28	11:02	11:10	11:15	11:25	11:34	---	11:44
							11:47	11:55	12:00	12:10	12:19	---	12:29
12:31	---	12:41	12:50	1:00	1:05	1:13							
1:16	---	1:26	1:35	1:45	1:50	1:58	12:32	12:40	12:45	12:55	1:04	---	1:14
2:01	---	2:11	2:20	2:30	2:35	2:43	1:17	1:25	1:30	1:40	1:49	---	1:59
2:46	---	2:56	3:05	3:15	3:20	3:28	2:02	2:10	2:15	2:25	2:34	---	2:44
							2:47	2:55	3:00	3:10	3:19	---	3:29
3:31	---	3:41	3:50	4:00	4:05	4:13							
4:16	---	4:26	4:35	4:45	4:50	4:58	3:32	3:40	3:45	3:55	4:04	---	4:14
5:16	---	5:26	5:35	5:45	5:50	---	4:32	4:40	4:45	4:55	5:04	---	5:14
6:16	---	6:26	6:35	6:45	6:50	6:58	5:32	5:40	5:45	5:55	6:04	---	6:14
---	7:42	7:44	7:51	8:01	8:04	8:11	7:13	7:20	7:23	8:32	7:40	7:42	---
---	8:42	8:44	8:51	9:01	9:04	9:11	8:13	8:20	8:23	8:32	8:40	8:42	---
---	9:42	9:44	9:51	10:01	10:04	---	9:13	9:20	9:23	9:32	9:40	9:42	---

On the "Eastbound" schedule, the times show when each bus leaves the Crossroads mall and arrives at each of its stops. The "Westbound" schedule shows when each bus will stop on its way to the Crossroads mall.

Example

What time does the 12:31 Eastbound bus arrive at the Cuming Stop?

Step 1. Find the departing time "12:31" in the Eastbound schedule under Crossroads. It is in the tenth row.

Step 2. Go straight across to the Cuming column. It is the fifth column.

Step 3. Read the time. 1:00

Answer: The 12:31 eastbound bus will arrive at Cuming at 1:00.

Solve each problem.

1. What time does the 6:16 Eastbound Crossroads bus reach the 61-Maple stop?

2. What time does the 8:01 Eastbound Crossroads bus reach the Burt stop?

3. What time does the Eastbound bus leave the Crossroads to reach the Jackson stop at 11:43?

4. What time does the Eastbound bus leave the Crossroads to reach the 90-Maple stop at 6:26?

5. To reach Crossroads at 9:29, what time does the Westbound bus stop at Cuming?

6. The 11:02 Jackson pickup by the Westbound bus arrives at the Crossroads at what time?

7. Brandy lives at 90th and Maple. She has an interview for a job at 23rd and Burt at 2:30 P.M. What time does she need to catch the bus at her home in order to make her interview on time?

8. Brandy's interview lasts one hour. What time will she get home if she takes the next available bus from 23rd and Burt?

9. The buses do not go to Crossroads after 6:14 P.M. What is the last pickup time at 90-Maple to get to Crossroads by 6:14 P.M.?

10. Jerry lives on Jackson Street and works at a mall called Crossroads. He must be at work at 9:00 A.M. What time will he need to catch the bus to Crossroads to be at work by 9:00 A.M.?

11. What time will he reach Crossroads?

12. How much time will the trip to work take?

13. Jerry gets off work at 6 P.M. What time will the Eastbound bus leave for Jackson Street?

14. What time will Jerry arrive at home?

15. If Jerry works late (after 6:30 P.M.), will he be able to take the Crossroads bus home?

16. The Westbound bus stops at 61-Maple at 3:55. What time does it arrive at Crossroads?

17. To reach Crossroads at 5:14, what time should you get on the Westbound bus at the 23-Burt stop?

18. On the Westbound bus, what would be the earliest and latest time that someone could arrive at the Crossroads stop?

Application: Trucking and Delivery Service

| United States Milage Chart For Selected Cities | | | | | | | | | | |
	Atlanta, GA	Chicago, IL	Dallas, TX	Denver, CO	Detroit, MI	Kanas City, MO	Los Angeles, CA	Louisville, KY	Memphis, TN	Milwaukee, WI	Minneapolis, MN
Atlanta, GA		708	822	1430	732	822	2191	415	382	799	1121
Chicago, IL	708		921	1021	279	542	2048	297	537	90	410
Dallas, TX	822	921		784	1156	505	1399	828	454	1015	949
Denver, CO	1430	1021	784		1283	606	1031	1119	1043	1038	920
Detroit, MI	732	279	1156	1283		769	2288	382	719	360	685
Kanas City, MO	822	542	505	606	769		1157	513	482	564	443
Los Angeles, CA	2191	2048	1399	1031	2288	1577		2182	1807	2069	1857
Louisville, KY	415	297	828	1119	382	513	2182		378	382	705
Memphis, TN	382	537	454	1043	719	482	1807	378		622	914
Milwaukee, WI	799	90	1015	1038	360	564	2069	382	622		337
Minneapolis, MN	1121	410	949	920	685	443	1857	705	914	337	

Use the formula $r = \dfrac{d}{t}$, where **r** is velocity, **d** is distance, and **t** is time. Use the mileage chart above to answer the following questions.

1. Joe drives a furniture truck around town. He spends 6 hours each day on the road and averages 45 mph. How many miles does he drive on an average day?

2. The distance between Altanta, GA, and Minneapolis, MN, is 1,121 miles. If a truck driver averages 60 mph, how long would it take her to travel between the two cities?

3. A truck driver picked up a load in Dallas, TX, and drove 8 hours to Kansas City, MO. What was the average velocity of the truck?

4. A long-distance hauler picked up a load of oranges in Los Angeles, CA, and drove to Denver, CO. Her average speed was 65 mph. She spent 6 hours sleeping during the trip and 3 hours on breaks and meals. How many hours did the trip take?

5. Jeremy drove a round trip, short-haul load between Chicago, IL, and Milwaukee, WI. Including a half-hour break, he was away from the office for 4 hours. What was his average driving speed?

6. Mary and Joe decided to move from Detroit, MI, to Los Angeles, CA. They packed their belongings into a rental van and left Detroit at 8:00 A.M. on Saturday. They drove straight through averaging 55 mph, including breaks and other stops. How long did it take them to drive to California? What time was it in California when they arrived? (Use the Time Zones chart on page 144.)

7. How long would it take Janice to drive from Louisville, KY, to Kansas City, MO, if she averages 50 mph?

8. A trucker carried a load from Dallas, TX, to Detroit, MI. If the actual driving time was 22 hours, what was the average velocity?

9. Jack wanted to fly from Denver, CO, to Atlanta, GA. If the plane averages 350 mph, how long is the flying time?

10. If Carla averages 60 mph on a drive between Milwaukee, WI, and Detroit, MI, how long will she be on the road?

Application: Speed of Planes

MILEAGE BETWEEN U.S. CITIES

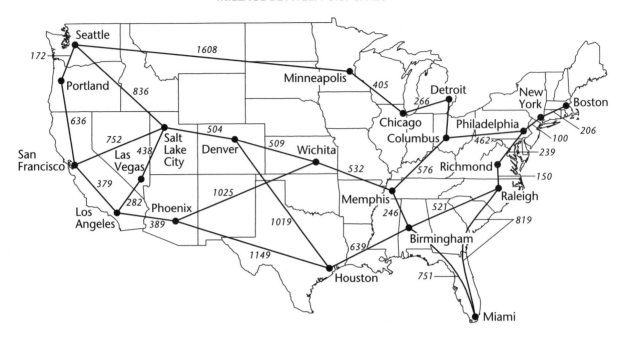

Use the formula $r = \dfrac{d}{t}$, where r is velocity, d is distance, and t is time. Use the mileage chart above to answer the following questions.

1. A jet picked up passengers in Denver, CO, and flew at an average velocity of 400 mph to Houston, TX. How long was the plane in the air?

2. A private plane picked up passengers in Denver, CO, and flew to Houston, TX. This plane averaged 125 mph. How long did the private plane take to reach Houston?

3. A jet flew approximately 2,000 miles from Seattle, WA, to Chicago, IL. The jet took 5 hours to make the trip. How fast was it flying?

4. The distance between New York City, NY, and Philadelphia, PA is 100 miles. How long would it take to fly between the two cities if the plane's speed is 125 mph?

5. Ann flew from Phoenix, AZ, to Wichita, KS in three hours. What was the speed of the plane?

6. Joe left Los Angeles, CA, for Seattle, WA. The plane flew at a speed of 365 miles per hour. How long did his flight take?

7. Joe took a smaller plane back home to Los Angeles from Seattle. He had a layover of one hour in Portland, OR, and his entire trip took seven hours. What was the average speed of the plane?

8. Heidi flew from Miami, FL, to Boston, MA. On the trip she stopped in Richmond, VA, where she had a two-hour layover. Her plane averaged 400 mph. How long did it take her to arrive in Boston?

9. If it takes $1\frac{1}{4}$ hours to fly from Columbus, OH, to Memphis, TN, what is the average velocity of the plane?

10. A helicopter made a special one-way flight from Chicago, IL, to Detroit, MI. If the trip took $2\frac{1}{2}$ hours, what was the speed of the helicopter? (Round your answer to the nearest whole number.)

11. A delivery plane started in Salt Lake City, UT, and then stopped in Denver, CO, and Wichita, KS, before ending the route in Memphis, TN. If the plane traveled about 200 mph, how long was the pilot in the air? (Round your answer to the nearest whole number.)

12. If it takes about 4 hours to fly from Minneapolis, MN, to Seattle, WA, what speed is the plane flying?

Section 5 Cumulative Review

Rewrite each measurement so it uses the given unit.

1. ____ seconds = 1 minute

2. 1 year = ____ months

3. ____ hours = 1 day

4. 1 month ≈ ____ weeks

5. 1 week = ____ days

6. ____ minutes = 1 hour

7. ____ weeks = 1 year

8. 1 month ≈ ____ days

9. 3 days = ____ hr

10. ____ min = $2\frac{1}{2}$ hr

11. 4 wk = ____ days

12. $4\frac{1}{2}$ yr = ____ mo

13. 150 sec = ____ min

14. 28 mo = ____ yr

15. 147 days = ____ wk

16. ____ hr = 270 min

Circle the longest time period.

17. 390 minutes; 5 hours; 4,200 seconds

18. 2 years; 110 weeks; 30 months

19. 18 days; 2 weeks; 300 hours

20. 6 months; 30 weeks; 1,080 hours

21. $2\frac{1}{2}$ minutes; 100 seconds; $\frac{1}{2}$ hour

22. 13 months; 1 year; 60 weeks

Fill in each blank with >, <, or =.

23. 6 years ____ 72 months

24. $4\frac{1}{2}$ days ____ 90 hours

25. 8 weeks ____ 60 days

26. 330 minutes ____ $6\frac{1}{2}$ hours

Mark the given duration on each stop watch.

27.   1 min 10 sec

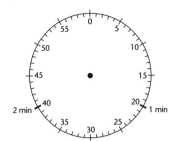

28.   2 min 42 sec

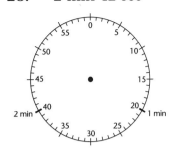

Add, subtract, multiply, or divide.

29. 6 hr 30 min – 2 hr 45 min

30. 3 days 18 hr × 5

31. 2 yr 7 mo + 3 yr 8 mo

32. 7 min 18 sec ÷ 6

33. 10 min 12 sec – 3 min 40 sec

34. 7 hr 45 min × 3

35. 6 wk 5 days + 3 wk 4 days

36. 5 days 3 hr 16 min ÷ 2

Circle the better estimate for each duration.

37. time to drink a 12-ounce soft drink   20 min or 20 sec

38. vacation time after 10 years on a job   15 wk or 15 days

39. time to microwave popcorn $2\frac{1}{2}$ sec or $2\frac{1}{2}$ min

40. time to sew a dress   17 hr or 17 days

41. time on a shift at a factory   8 days or 8 hr

Solve each problem.

42. George has 20 days of sick leave accumulated at his workplace. He must have surgery, and will be off work for 6 weeks. Does he have enough sick days for his recovery period? (Assume he works a 5-day work week.)

43. Henri works these hours during one week: 7 hours 50 minutes; 8 hours 20 minutes; 8 hours 10 minutes; 7 hours 45 minutes; and 8 hours 55 minutes. What was his average daily work time?

44. Carol exercises 1 hour 20 minutes daily. If she does this 6 days each week, how much time does she spend each week exercising?

45. At Josie's work site, she is supposed to have a 15-minute break every 3 hours. She has worked 200 minutes since her last break. Has she worked long enough to earn the 15-minute break?

Solve each of the following problems about time periods and health medications.

46. Joe has high blood pressure. He must take a pill twice a day to control his blood pressure. How many hours apart should he take his pills?

47. Joe took his first pill of the day at 6:15 P.M. When should he take his second pill?

48. Ann had to take a pain killer every four hours after her wisdom teeth were removed. If she took her first pill at 2:30 P.M., what times would she need to take the next three pills?

49. Mary took two aspirins for a headache. The aspirin bottle warned not to take more than 12 tablets in a 24-hour period. How often could she take a pair of aspirin during a 24-hour period?

50. If she took her first two aspirin at 7:00 A.M., at what times could she take another two aspirins over the next 24 hours?

51. Jess has received an antibiotic for his bronchitis. The instructions says to take four capsules, three times a day. If he takes his first set of four capsules at 6:45 A.M., what time should he take his next set of capsules?

Find the volume of each container.

52.

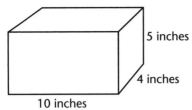

5 inches
4 inches
10 inches

55.

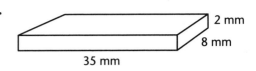

2 mm
8 mm
35 mm

53.

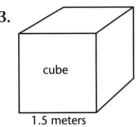

cube

1.5 meters

56.

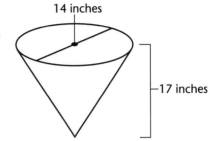

14 inches

17 inches

54.

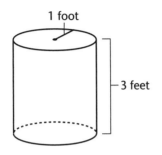

1 foot

3 feet

57.

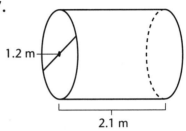

1.2 m

2.1 m

Find each indicated length.

58.

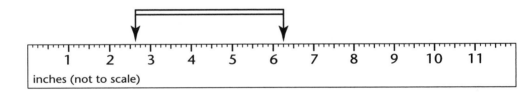

1   2   3   4   5   6   7   8   9   10   11

inches (not to scale)

59.

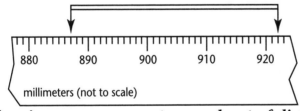

880   890   900   910   920

millimeters (not to scale)

Mark the given measurement on each set of dials.

60. 25,033

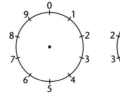 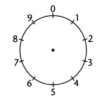

61. 40,995

  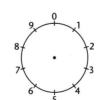 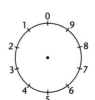 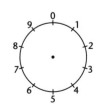

Post-Test
Measurement Skills Review

The purpose of this review is to see how well you have mastered the skills in this measurement book. Take your time and work each problem carefully.

Rewrite each measurement so it uses the given unit.

1. 1 lb = _____ oz

2. 2.5 T = _____ lb

3. 3,500 mg = _____ g

4. 4.7 cm = _____ mm

5. 84 in. = _____ yd

6. 1 mi = _____ ft

7. 3.2 kL = _____ L

8. $6\frac{1}{2}$ qt = _____ c

9. 475 mL = _____ L

10. $3\frac{1}{4}$ hr = _____ min

11. 1 yr = _____ days

12. 287 days = _____ wk

Write the amount shown by each scale.

13.

14.

15.

16.

17.

Add, subtract, multiply, or divide.

18. 2 lb 4 oz + 3 lb 11 oz =

19. 7 lb 2 oz – 3 lb 7 oz =

20. 10 g + 400 mg =

21. 4.8 cm × 7 =

22. 1,250 km ÷ 5 =

23. 6 yd 2 ft + 4 yd 2 ft =

24. 5 qt 1 c – 2 qt 3 c =

25. $6\frac{1}{2}$ gal × 4 =

26. 275 cm ÷ 2.5 =

27. 4 days 18 hours + 2 days 20 hours =

28. 7 weeks 5 days ÷ 9 =

29. 12 hr 14 min – 7 hr 36 min =

30. 4 ft 3 in. × 10

Questions 31–33.

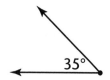

31. What kind of angle is this?

_____

32. What is the measure of its complement?

_____

33. What is the measure of its supplement?

_____

Questions 34–35

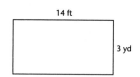

14 ft

3 yd

34. What is the perimeter?

_____

35. What is the area?

_____

36. Jose was going to make a circular flower bed. If the diameter of the circle is 10 feet, what is the circumference of the flower bed?

37. What is the area of the circular flower bed?

38. If he wants to add 6 inches of top soil, what is the volume of soil needed?

39. Randy's daughter had a fever. The thermometer said 102.4°F. How many degrees above 98.6°F was the child's fever?

40. After having the flu for a few days, Carla found she had lost $6\frac{1}{2}$ pounds. If she now weighs 134.8 pounds, how much did she weigh before she had the flu?

41. Ryan works as a lineman for a cable television company. If he cut $38\frac{1}{2}$ yards of cable from a new 200-yard spool, how much cable is left on the spool?

42. Chris and Hank are marathon runners. Chris entered a 10-mile run. Hank entered a 10-kilometer run. Who ran farther?

43. Tran used $22\frac{1}{2}$ centimeters of copper wire in each appliance he repaired. If he fixed eight appliances, how many meters of copper wire did he use?

44. There are $6\frac{1}{2}$ grams of fat in a blueberry muffin. If Jeni ate 2 muffins each morning for four days in a row, how many grams of fat did she eat?

45. At the local art museum, a lawn display includes a large cement cone. If the length of the cone is 12 meters and its diameter is 8 meters, what is the volume of the cement cone?

46. What is the volume of a cube with sides 10 feet long?

47. Manual worked from 7:15 A.M. to 12:30 P.M. on Monday and from 7:30 A.M. to 1:45 P.M. on Tuesday. If he earns $7.20 an hour, how much did he earn for his work on Monday and Tuesday?

48. Carl's odometer reading at the start of a trip was 28,472.6 miles. After driving a full day, the odometer read 28,875.2 miles. How far had he driven? If he had driven for 6 hours, what was his average rate?

49. A cargo plane averages 11.6 miles per gallon of fuel. How many gallons of fuel are needed for a trip of 1,485 miles? (Round your answer to the nearest gallon.)

50. Kai is paid time-and-a-half for each hour over 40 hours that she works in a week. If she works $8\frac{1}{2}$ hours each day for five days, and her base pay is $6.40 per hour, how much overtime pay will she earn?

Answers are on page 162.

Measurement Skills Review Answers

1. 16
2. 5,000
3. 3.5
4. 47
5. $2\frac{1}{3}$
6. 5,280
7. 3,200
8. 26
9. 0.475
10. 195
11. 365
12. 41
13. $2\frac{3}{4}$ lb
14. $31\frac{7}{8}$ in.
15. $1\frac{1}{4}$ c
16. 84°F
17. 38 min

18. 5 lb 15 oz
19. 3 lb 11 oz
20. 10.4 g or 10,400 mg
21. 33.6 cm
22. 250 km
23. 11 yd 1 ft
24. 2 qt 2 c
25. 26 gal
26. 110 cm
27. 7 days 14 hr
28. 6 days
29. 4 hr 38 min
30. 42 ft 6 in.
31. acute
32. 55°
33. 145°
34. 46 ft or $15\frac{1}{3}$ yd

35. 14 sq yd or 126 sq ft
36. 31.4 ft
37. 78.5 sq ft
38. 39.25 cu ft
39. 3.8°
40. 141.3 lb
41. 161.5 yd
42. Chris
43. 1.8 m
44. 52 g
45. 200.96 cu m
46. 1,000 cu ft
47. $82.80
48. 402.6 miles, 67.1 mph
49. 128 gal
50. $24

Skills Review Chart

If you had fewer than 40 correct, review the sections where you missed problems. Rework any problems you missed. Here is a list of the problems and where each skill is covered.

Problem Numbers	Section
4–6, 14, 21–23, 26, 31–37, 41–43	Section 2: Length and Angles
1–3, 13, 16, 18–20, 30, 39, 40, 44	Section 3: Weight and Temperature
7–9, 15, 24, 25, 38, 45, 46	Section 4: Capacity and Volume
10–12, 17, 27–29, 47–50	Section 5: Time and Velocity

Formulas

Area
Rectangle $\quad A = \ell \times w$
Square $\qquad A = s^2$
Circle $\qquad A = \pi\, r^2$

Volume
Box $\qquad V = \ell \times w \times h$
Cube $\qquad V = s^3$
Cylinder $\quad V = \pi\, r^2 h$
Cone $\qquad V = \frac{1}{3}\pi\, r^2 h$

Circumference of a circle
$C = \pi d$ or $C = 2\pi r$

Temperature
$C = \frac{5}{9}(F - 32) \qquad F = \frac{9}{5}C + 32$

Measurement Units

Celsius	Temperature	Fahrenheit
0°C	water freezes	32°F
100°C	water boils	212°F
37°C	normal body temperature	98.6°F

Time
60 seconds (sec) = 1 minute (min)
60 min = 1 hour (hr)
12 months (mo) = 1 yr

24 hr = 1 day
7 days = 1 week (wk)
52 wk = 1 year (yr)

Metric

Weight
1,000 milligrams (mg) = 1 gram (g)
1,000 g = 1 kilogram (kg)

Customary
16 ounces (oz) = 1 pound (lb)
2,000 lb = 1 ton (T)

Length
1,000 millimeters (mm) = 1 meter (m)
100 centimeters (cm) = 1 m
1,000 m = 1 kilometer (km)

12 inches (in.) = 1 foot (ft)
3 ft = 1 yard (yd)
36 in. = 1 yd
5,280 ft = 1 mile (mi)
1,760 yd = 1 mi

Capacity
1,000 milliliters (mL) = 1 liter (L)
1 kiloliter (kL) = 1,000 L

3 teaspoons (tsp) = 1 tablespoon (tbsp)
1 fluid ounce (fl oz) = 2 tbsp
8 fl oz = 1 cup (c)
2 c = 1 pint (pt)
2 pt = 1 quart (qt)
4 qt = 1 gallon (gal)

Glossary

angle
a figure formed by two lines that start at the same point. The point is called the **vertex** of the angle and the lines are called the **sides.**

Angles can be measured in **degrees.** An **acute angle** is smaller than 90°, a **right angle** is exactly 90°, an **obtuse angle** is between 90° and 180°, and a **straight angle** is exactly 180°.

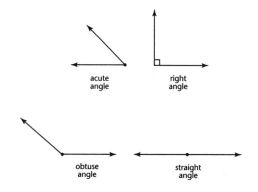

area
the physical amount of a region covered by a flat figure. Area can be measured directly, by covering a figure with squares, or area can be calculated based on lengths of parts of the figure.

base
of a cone or cylinder; the side or sides of a cone or cylinder that are circles.

box
a container whose top, bottom, and sides are rectangles.

capacity
the physical amount that an object can hold or contain.

Celsius
a temperature scale. On a Celsius scale, the freezing point of water is 0°C and the boiling point of water is 100°C.

centi-
a prefix used with metric units; it means one-hundredth.

circle
a round, flat figure. Every point on a circle is the same distance from the **center** of the circle. A line segment from the center to any point on the circle is a **radius**, and a line segment containing two points on the circle <u>and</u> the center of the circle is a **diameter.**

circumference
the distance around a circle.

compare and order
tell whether two objects are the same or different ("compare"), and then tell which one is greater than or less than the other ("order").

complementary angles
two angles whose measures add up to 90°.

cube
a box-shaped object whose six sides are squares.

cylinder
an object shaped like a tube or a rod; it has circular ends that are the same size.

degree
a unit for measuring angles; the symbol is °. There are 360 degrees in a full circle. (It is also a unit for measuring heat.)

diagonal
a line segment that connects two corners of a figure that are separated by at least one other corner of the figure.

dimensions
a term used to refer to the length and width of a rectangle or to the length, width, and height of a box.

distance
the physical amount of length between two points or objects.

duration
another term for "amount of time" or "time interval."

equivalent measurement

two measurements that represent the same amount. For example, the three measurements 36 inches, 3 feet, and 1 yard are equivalent.

estimate

as an object or noun: an approximate value; as an action or verb: to calculate or make an educated guess for an approximate value.

Fahrenheit

a temperature scale. On a Fahrenheit scale, the freezing point of water is 32°F and the boiling point of water is 212°F.

height

a physical amount of distance, measured vertically or up-and-down.

kilo-

a prefix used with metric units; it means one thousand.

length

a physical amount of distance.

level

a carpenter's term for horizontal.

measurement

a number and a unit. Also, as a general term *measurement* means using tools to assign numbers and units to objects, and then using those numbers and units to compare and relate objects.

milli-

a prefix used with metric units; it means one-thousandth.

operation

a term used to refer to addition, subtraction, multiplication, or division.

parallel lines

two or more lines that go in the same direction and never meet. When two parallel lines are crossed by a third line, they form four pairs of **corresponding angles** and four pairs of **alternate interior angles**.

perimeter

the distance around a figure.

plumb

a carpenter's term for vertical or straight up-and-down.

protractor

a tool used for measuring angles. It usually has numbers 0 through 180 marked along a half-circle.

rate

another term for velocity or speed. In general, rate can refer to any fraction where the numerator is a measurement and the denominator is a measurement that uses the number 1.

ruler

a tool for measuring length.

scale

any tool for measuring, especially for measuring weight.

scale drawing

a drawing of an object where the lengths in the drawing represent lengths on the actual object.

scale on a map

a key telling how the lengths on the map represent actual distances.

semicircle

half a circle.

sphere

a ball-shaped object.

supplementary angles

two angles whose sum is 180°.

temperature

the physical feeling of hot and cold.

thermometer

a tool for measuring temperature.

thermostat

a device for regulating temperature. It uses a thermometer to tell when to turn a heating or cooling system on and off.

time

measured in seconds, minutes, hours, and so on. The tool used for measuring a *duration of time* is the stopwatch; the tool used for recording time is the watch or clock.

time zone
> one of 24 regions of the Earth; as you cross a time zone you change by one hour the time showing on your clocks.

unit
> the amount of length, capacity, or velocity that represents one inch, one cup, one mile per hour, and so so. **Customary units** include inches, pounds, and quarts; **metric units** include meters, grams, and liters.

velocity
> how fast an object is traveling. Usually, velocity is not measured directly, but is calculated based on traveling a particular distance in a particular time.

volume
> the physical amount of space used by an object. Usually, volume is not measured directly but is calculated based on lengths of parts of the object.

weight
> the physical feeling of how heavy or light an object is. The tool for measuring weight is called a scale.

zero mark
> the starting place on a ruler or scale.

Index